Transcription Factors CREB and NF-κB:
Involvement in Synaptic Plasticity and Memory Formation

Edited By

Benedict C. Albensi, *Ph.D.*

University of Manitoba
Canada

eBooks End User License Agreement

DEDICATION

This e-Book is dedicated to all those that suffer from memory impairments – may this research contribute to improved therapies for these individuals

CONTENTS

CHAPTERS

About the Editor

Dr. Albensi completed his PhD in Neuroscience at the University of Utah's School of Medicine in 1995. He obtained advanced training as a Postdoctoral Fellow at the Georgetown Institute of Cognitive and Computational Sciences (GICCS) at Georgetown University Medical Center where he worked with a high field MRI system for assessing brain trauma and its effect on hippocampal structure and physiology. Thereafter he trained under Dr. Mark Mattson as a Postdoctoral Scholar at the Sanders-Brown Center on Aging at the University of Kentucky's Medical School and studied the involvement of NF-kB in synaptic plasticity. Subsequently, he participated in clinical trials research at Pfizer, Inc. in Ann Arbor, Michigan. Currently he holds the Honourable Douglas Everett, Patricia Everett and the Royal Canadian Properties Endowment Fund Chair and is an Associate Professor of Pharmacology and Therapeutics at the University of Manitoba, Winnipeg, Canada. In addition, he is an Adjunct Professor of Electrical and Computer Engineering at this same school and a Principal Investigator in the Division of Neurodegenerative Disorders at the St. Boniface Research Centre. He is also a Research Affiliate for the University of Manitoba's Centre on Aging and a Scientist at the Manitoba Institute for Child Health (MICH). He has published more than 100 articles that include peer-reviewed experimental papers, invited book chapters and reviews, abstracts, and other CNS relevant manuscripts on synaptic plasticity, memory, and neuronal excitability. In addition he has been a member of several grant review panels around the world, including the NIH, CIHR, Alzheimer's Society of Canada, Alzheimer's Association (USA), Wellcome Trust, American Institute of Biological Sciences, and the U.S. Civilian Research & Development Foundation. He is also on the editorial board of CNS and Neurological Disorders – Drug Targets. He is married and lives in Winnipeg, Canada with his four daughters.

FOREWORD

The world was originally introduced to the concept of synaptic plasticity over 60 years ago, when Dr. Donald Hebb first clearly defined a physiological mechanism for learning and memory in his seminal work *"The Organization of Behavior"*. It took another 20 years for Bliss and Lomo to scientifically validate Hebb's postulate, and show that neurons could alter their ability to communicate with one another in a persistent manner. Together, these works started off what has grown to become the field of synaptic plasticity. The years following the initial discovery were exciting times for learning and memory young researchers like myself, and each discovery over the next 20 years seemed to push us closer to elucidating the biological mechanisms responsible for memory formation. This seemed particularly true in the mid-1980's when the NMDA receptor was being heralded as the key to learning and memory processes. However, more recently it has become obvious that the activation of membrane receptors is only the first step in a cascade of post-synaptic events that ultimately results in genomic changes. While it remains unclear whether memories reside in our genes, it is clear that several gene transcription factors play a role in determining how easily and reliably changes in neuronal communication can be established. The works included in this e-Book help to explain the evidence for two specific transcription factors, cAMP response element-binding (CREB) and NF-kB, being involved in synaptic plasticity. Furthermore, recent evidence is presented for how alterations in the normal functioning of these transcription factors can play a role is some specific disease processes. These topics included are essential reading for any student of the mind, and particularly for those of us engaged in synaptic plasticity research.

Brian R. Christie

Division of Medical Sciences, University of Victoria, Victoria, BC
The Island Medical Program, University of British Columbia, Victoria, BC

PREFACE

Transcription factors are specialized proteins that bind to specific DNA sequences, thereby controlling the movement (a.k.a. transcription) of genetic information from DNA to mRNA. In so doing, transcription factors can promote or repress the recruitment of RNA polymerase to specific genes thereby regulating gene expression.

Synaptic plasticity is the capacity of the connections between neurons to change in strength in response to alterations in synaptic transmission and other cellular signals. So-called plastic change is possible as a result of several mechanisms that function co-operatively in the synaptic region and results in morphological change. The formation of memory is theorized to be associated with and driven by mechanisms of synaptic plasticity.

The main aim of the authors for this e-Book was to critically survey the role of two recognized transcription factors in processes of synaptic plasticity and memory. In addition, the authors provided recent data from their own labs and in some cases provided a perspective relevant to specific CNS diseases and potential drug targets. Historically, the transcription factor cAMP response element-binding (CREB) has been the most well documented transcription factor shown to play a role in synaptic plasticity and memory. CREB has several functions, but its most notable function has to do with the formation of long-term memories where knock out of this protein not only impairs memory, but also has a negative effect on cell survival. More recently, other transcription factors, such as CCAAT/enhancer binding protein (C/EBP), early growth response (Egr) protein, activator protein 1 (AP-1), and nuclear factor kappa B (NF-κB) have been implicated in synaptic plasticity and memory as well. Of these, scientific literature on NF-κB's theorized role in synaptic plasticity and memory is growing rapidly. Interestingly, in some recent studies CREB and NF-κB have also been shown to interact with each other where both contribute in a cooperative fashion to the initiation of gene expression.

This e-Book is divided into seven chapters. In the first two chapters, CREB is reviewed and roles for this protein are evaluated not only in normal long term memory, but also in a context of memory impairment in disease states such as Alzheimer's disease (chapter one), Rubinstein-Taybi syndrome (chapter one), and some neuropsychiatric disorders (chapter two), etc. The third chapter serves as a bridge for our two highlighted transcription factors, where the author presents a synthesized discussion of the literature on CREB and NF-κB and also on other transcription factors that are relatively unrecognized at the time of this writing. In the fourth chapter, the potential role of NF-κB in synaptic plasticity and memory is discussed and recent evidence for its participation from several models is presented. In chapter five, a more specific role for NF-κB is considered that involves adult neurogenesis. The sixth chapter discusses NF-κB and specific mechanisms of transcriptional regulation and puts into perspective how NF-κB may play a role in several forms of memory including, memory consolidation, reconsolidation and extinction.

The final chapter serves well not only to summarize and question NF-κB's role in synaptic plasticity and memory, but also to point out important technical considerations that may have influenced (rightly or wrongly) many past studies in this field.

We are living in an exciting time for memory research. The last several decades have seen an explosion in scientific data and literature that have contributed to our understanding of the neurological basis of memory. In spite of this plethora of information many outstanding questions still remain: What processes underlie the formation of new memories and the establishment of long term memory? What mechanisms determine the strength of memory? Which molecules play essential roles in these processes? How do the transcription factors CREB and NF-κB contribute to processes of long term memory? The chapters in this e-Book attempt to address some of these most interesting, but unanswered questions.

Benedict C. Albensi, Ph.D.
University of Manitoba
Canada

List of Contributors

Benedict C. Albensi *Division of Neurodegenerative Disorders, St. Boniface General Hospital Research Centre, 351 Tache Ave, St. Boniface Research Centre, R4050, Winnipeg, MB, R2H 2A6 Canada and Department of Pharmacology and Therapeutics, University of Manitoba, Winnipeg, MB, A203 Chown Bldg., 753 McDermot Avenue, Winnipeg, MB, R3T 2N2 Canada*

Steven W. Barger *Department of Geriatrics, Department of Neurobiology and Developmental Sciences, University of Arkansas for Medical Sciences, Little Rock AR, 72205 USA and Geriatric Research Education and Clinical Center, Central Arkansas Veterans Healthcare System, Little Rock AR 72205 USA*

Ángel M. Carrión *División de Neurociencias, Universidad Pablo de Olavide, Carretera de Utrera Km. 1, Sevilla-41013, Spain*

Ángela Fontán-Lozano *División de Neurociencias, Universidad Pablo de Olavide, Carretera de Utrera Km. 1, Sevilla-41013, Spain*

Mariagrazia Grilli *DiSCAFF, University of Piemonte Orientale "A. Avogadro" & DFB Center, Laboratory of Neuroplasticity and Pain, Novara 28100 Italy*

Sheena A. Josselyn *Program in Neurosciences & Mental Health, Hospital for Sick Children Research Institute, 555 University Ave. Toronto ON, M5G 1X8 Canada and Department of Physiology, University of Toronto, 1 King's College Circle, Toronto, ON M5S 1A8 Canada*

Xianrong R. Mao *Department of Anesthesiology, Washington University School of Medicine, St. Louis MO, 63110 USA*

Vasco Meneghini *DiSCAFF, University of Piemonte Orientale "A. Avogadro" & DFB Center, Laboratory of Neuroplasticity & Pain, Novara 28100 Italy*

Keith J. Murphy *Applied Neurotherapeutics Research Group, UCD School of Biomolecular and Biomedical Science, Conway Institute, University College Dublin, Belfield, Dublin 4, Ireland*

Gary Odero *Division of Neurodegenerative Disorders, St. Boniface General Hospital Research Centre, 351 Tache Ave, St. Boniface Research Centre, Winnipeg MB, R2H 2A6 Canada*

Niamh C. O'Sullivan *Applied Neurotherapeutics Research Group, UCD School of Biomolecular and Biomedical Science, Conway Institute, University College Dublin, Belfield, Dublin 4, Ireland*

Eva M. Pérez-Villegas *División de Neurociencias, Universidad Pablo de Olavide, Carretera de Utrera Km. 1, Sevilla-41013, Spain*

Asim J. Rashid *Program in Neurosciences & Mental Health, Hospital for Sick Children Research Institute, 555 University Ave. Toronto ON, M5G 1X8 Canada*

Arturo Romano *Laboratorio de Neurobiología de la Memoria, Departamento de Fisiología, Biología Molecular y Celular, Facultad de Ciencias Exactas y Naturales, Universidad de Buenos Aires, IFIBYNECONICET, Ciudad Universitaria, Pab. II, 2do Piso (1428) Buenos Aires, Argentina*

Rocío Romero-Granados *División de Neurociencias, Universidad Pablo de Olavide, Carretera de Utrera Km. 1, Sevilla-41013, Spain*

Graham K. Sheridan   *Applied Neurotherapeutics Research Group, UCD School of Biomolecular and Biomedical Science, Conway Institute, University College Dublin, Belfield, Dublin 4, Ireland*

Wanda Snow   *Division of Neurodegenerative Disorders, St. Boniface General Hospital Research Centre, 351 Tache Ave, St. Boniface Research Centre, R4050, Winnipeg, MB, R2H 2A6 Canada*

Kunjumon Vadakkan   *Division of Neurodegenerative Disorders, St. Boniface General Hospital Research Centre, 351 Tache Ave, St. Boniface Research Centre, R4050, Winnipeg, MB, R2H 2A6 Canada*

CHAPTER 1

Regulation of Synaptic Plasticity and Long-Term Memory by CREB: Implications for Targeting Memory Disorders Including Alzheimer's Disease and Rubinstein-Taybi Syndrome

Asim J. Rashid and Sheena A. Josselyn*

Program in Neurosciences & Mental Health, Hospital for Sick Children, Research Institute, 555 University Ave. Toronto, ON, M5G 1X8 Canada, Department of Physiology, University of Toronto, Toronto ON, Canada and Institute of Medical Sciences, University of Toronto, Toronto, ON, M5G 1X8, Canada

Abstract: The cAMP Responsive Element Binding Protein (CREB) is an activity regulated transcription factor that modulates the transcription of genes with cAMP responsive elements (CRE) located in their promoter regions. In the central nervous system (CNS), an abundance of evidence has indicated that CREB-mediated transcription can be triggered through a number of pathways and is central to the induction of long-term synaptic changes and the formation of long-term memory. Dysfunction of CREB signaling has been implicated in the memory/cognitive impairments associated with a variety of CNS disorders, including Rubinstein-Tabyi Syndrome and Alzheimer's disease, suggesting that CREB or proteins associated with CREB-mediated transcription may be a useful therapeutic target in the treatment of these disorders. Here, we review the key features of CREB-dependent transcription, the role for CREB in synaptic plasticity and long-term memory formation, and also summarize the data implicating CREB dysfunction in some of the disorders in which the formation and storage of memories are affected.

Keywords: CREB, Intracellular signaling, Synaptic plasticity, Long-term potentiation, Long-term memory, Fear memory, Spatial memory, Transgenic mice, Alzheimer's disease, Rubinstein-Taybi syndrome, Coffin-Lowry syndrome.

INTRODUCTION

The ability of an organism to adapt and respond to its environment depends largely on the ability of that organism to encode and store salient information from previous experiences and retrieve that information for use in an appropriate context. In higher organisms, these processes, namely learning, memory storage and memory retrieval, are critical for proper cognitive function. Disruptions in these processes have been implicated in age-related cognitive decline as well as in the pathophysiology of a variety of human disorders such as Rubinstein-Taybi syndrome and Alzheimer's disease, among others. Determining the mechanisms by which these events occur, therefore, is a central goal of neuroscience research.

An extensive amount of information has already been obtained about the mechanisms by which memories are formed and stored in the central nervous system (CNS). It is generally understood that memories can last for dramatically different lengths of time, from seconds or minutes (short-term memory, STM) to days, months or years (long-term memory, LTM). Depending on the nature of the memory, STM and LTM may be encoded in subsets of neurons in functionally discrete areas of the brain or represented in networks of neurons in different interacting regions [1, 2]. At the molecular level, STM is thought to be mediated by covalent modifications of existing synaptic molecules, such as phosphorylation or dephosphorylation of enzymes, receptors or ion channels [3]. In contrast,

***Address correspondence to Sheena A. Josselyn:** Program in Neurosciences & Mental Health, Hospital for Sick Children, Research Institute, 555 University Ave. Toronto, ON, Canada M5G 1X8; E-mail: sheena.josselyn@sickkids.ca

LTM persists for days or longer, and is thought to be mediated by growth of new synapses and long-term plastic changes of existing synapses [4]. Notably, a wealth of evidence from evolutionarily divergent species indicates that, unlike STM, LTM requires the transcription and translation of new proteins for these long-term synaptic changes [5, 6]. Therefore, the dynamic regulatory mechanisms that direct transcription in the nervous system plays a critical role in the long-term storage of information.

One of the most extensively studied transcription factors in the formation of LTM is cAMP (cyclic adenosine 3',5'-monophosphate) - response element binding protein (CREB). CREB proteins are a family of transcription factors that modulate the transcription of genes by binding to cAMP responsive elements (CRE) in their promoter regions. The majority of the studies examining the role of CREB in memory have indicated a prominent role in the long-term storage of various types of memories in different regions of the mammalian brain. CREB has also been shown to be critical for the formation of LTM in the simpler nervous systems of fruit flies and snails. The importance of CREB signaling in brain function is further indicated by evidence that CREB signaling may be disrupted in various CNS disorders and these disruptions correlate with aspects of cognitive dysfunction associated with these disease states. Here we will review the evidence supporting the role for CREB in LTM and synaptic plasticity, which is the putative molecular correlate of LTM formation. We will also review the biochemical basis for CREB action and discuss the possibility of targeting the CREB signaling pathway for the treatment of disorders in which memory processes are disrupted.

STRUCTURE OF CREB AND MECHANISM OF CREB SIGNALING

The CREB family of transcription factors originates from three genes: CREB, CREM (cAMP Response Element Modulator) and ATF-1 (Activating Transcription Factor 1) [7-9]. The CREB gene encodes three alternatively spliced variants (CREB α, β and σ) each of which, along with ATF-1 and CREMτ and CREMτα splice variants, can stimulate transcription through promoters containing CRE sequences. In contrast, the four major splice variants of CREM in the brain (CREM α, β, γ and ICER [Inducible cAMP Early Repressor]) have been shown to repress CRE-dependent transcription.

CREB, CREM and ATF1 proteins share a conserved basic leucine zipper (bZip) domain that is responsible for dimerization between CREB family members as well as DNA binding [10]. The transcriptional activators, but not repressors, in the CREB family also contain two glutamine-rich constitutive activation domains (Q1 and Q2) that flank a kinase-inducible transactivation domain (KID). Activation of CREB-mediated transcription is promoted by phosphorylation of CREB at Ser133 in the KID domain (for review see [11]). Several kinases have been shown to phosphorylate CREB at Ser133, typically in response to increases in intracellular calcium and/or cAMP, and include protein kinase A (PKA), mitogen-activated protein kinase (MAPK)-activated ribosomal S6 kinase (RSK), calcium-calmodulin-dependent kinases I and IV (CaMKI, CaMKIV) and mitogen and stress-activated kinase 1 (MSK1) [12-23] (Fig. **1**). These kinases, therefore, serve to transmit channel and/or receptor generated signals from the plasma membrane to CREB in the nucleus. The importance of Ser133 phosphorylation for CREB signaling has been demonstrated by the elimination of stimulus-induced transcriptional activation by mutation of this site to a non-phosphorylatable alanine [24].

CREB can be bound to CRE independently of its phosphorylation state at serine 133 (S133). Phosphorylation of CREB at S133 promotes binding of the CBP/p300 complex, which serves to recruit basal transcription machinery. TORC can translocate to the nucleus in response to intracellular increases in calcium and/or cAMP and enhance transcription through its interactions with CREB as well as CBP. A large number of kinases and phosphatases, such as those shown here, can positively or negatively modulate the functions of CREB, CBP and TORC, underlying multiple levels of cellular regulation of transcription from CRE and therefore multiple potential therapeutic targets. For simplicity, other proteins important for CREB-mediated transcription, such as TFIID, have been omitted. (see text for abbreviations)

The occupancy of CRE elements by CREB is not sufficient for activation of transcription and appears to be independent of the CREB phosphorylation state [25, 26]. Binding of CREB to CRE is likely determined by the methylation status of histones at specific promoters [27], which may be subject to regulation in a tissue and/or cell specific manner.

Phosophorylation of CREB at Ser133, rather, promotes binding to transcriptional co-activators CREB binding protein (CBP) and p300, which support recruitment of basal transcription machinery to the CRE-containing transcript and also acetylate histones to facilitate transcription. A recent report also has suggested that phosphorylation of CREB at Ser133 prevents its degradation thereby further promoting transcriptional activation [28].

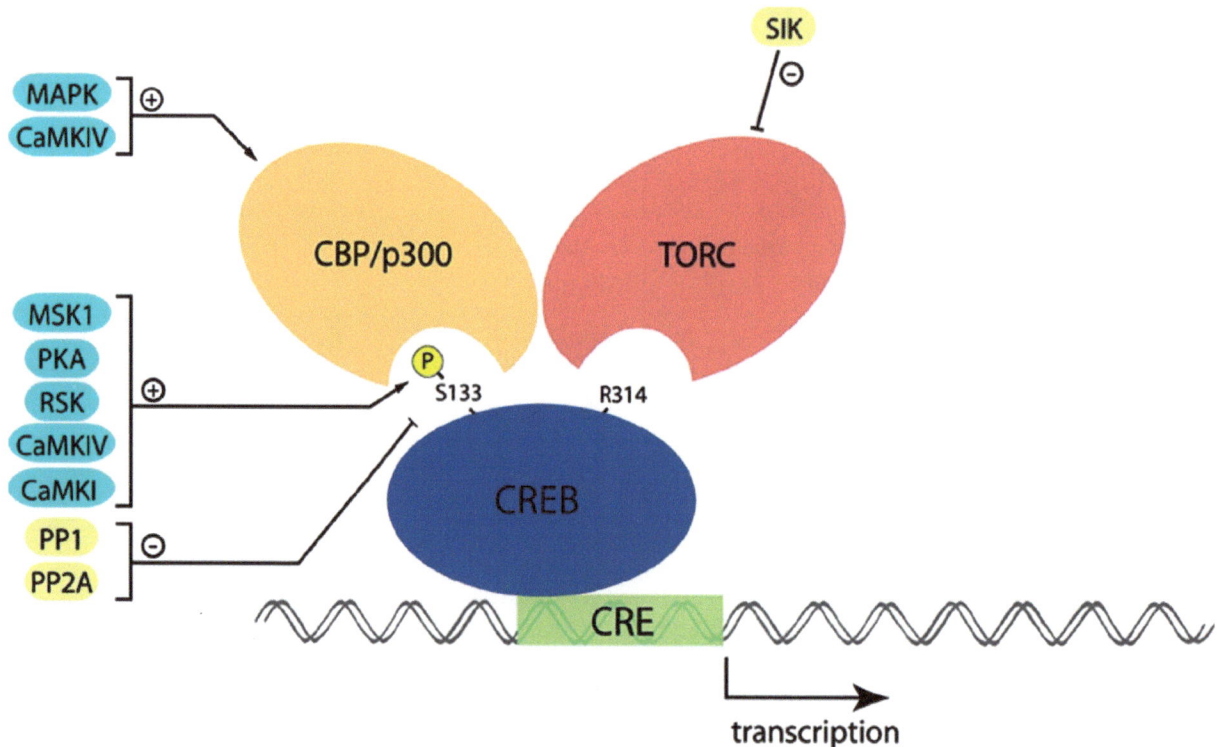

Fig. (1). Schematic of CREB-mediated transcription.

Although phosphorylation of CREB and subsequent recruitment of CBP/p300 are critical factors for cAMP-mediated transcription, it had been noted in studies that phosphorylation of CREB did not completely correlate with transcription of CRE-containing genes and the time course of CREB phosphorylation often outlasted the time window in which CREB-dependent transcription occurs. These observations implicated the necessity for an additional factor critical for the initiation and time course of CREB-mediated transcription. Accordingly, a family of three proteins known as TORCs (Transducers of CREB Regulated Activity) has recently been identified that can translocate from the cytosol to the nucleus in response to elevations in both intracellular cAMP and calcium and bind to Arg314 in the bZip domain of CREB, independent of the CREB phosphorylation state (see [29] for review). The TORC1 isoform is highly expressed in the rodent hippocampus and has been shown to stimulate CREB-dependent transcription and enhance synaptic plasticity [30, 31].

Along with CREB, CBP/p300 and TORC, a number of other important components have also been identified that are part of, or which can modulate, the CRE-dependent transcription apparatus. For example, TFIID (transcription factor IID) binds to the Q2 domain of CREB and is required for CREB transactivation, while ATF-4 can repress CREB mediated activity in neurons and is subject to dynamic regulation by the transcription initiation factor eIFα2. The involvement of multiple proteins for CREB-mediated transcription, and the potential to utilize different functional isoforms of some of them, underlies the complexity and diversity of transcriptional mechanisms from CRE promoters. This diversity may indicate the use of specific combinations of required proteins to reflect cell-, signal- or gene-specificity of CREB-mediated transcription [32].

The involvement of multiple proteins in CREB signaling also means that multiple targets exist for dynamic regulation of transcription by the cell. Furthermore, these points of regulation could be considered possible therapeutic targets in diseases that involve disruptions of CREB signaling. As mentioned above, Ser133 phosphorylation of CREB can occur through the action of several possible kinases. Similarly, phosphatases such as protein phosphatase 1 (PP1) and PP2A can dephosphorylate Ser133 in response to certain cellular signals [13, 18, 33]. CBP activity is also regulated by phosphorylation, specifically by MAPK and CaMKIV [34, 35], while TORC2 can be activated by increases in intracellular cAMP which prevents inhibitory phosphorylation of TORC2 by salt-inducible kinase 2 (SIK-2) [36].

CREB AND SYNAPTIC PLASTICITY

Long-term changes in synaptic transmission, referred to as long-term potentiation (LTP) and long-term depression (LTD), have been proposed to be the cellular correlates of LTM formation in the brain. Repetitive stimulation of neurons *in vitro* at high (>100Hz) or low (~1Hz) frequencies results in enhancement (LTP) or attenuation (LTD) of synaptic transmission, respectively, that can last at least several hours. Although the link between LTP/LTD and LTM has not been unequivocally proven, there are striking parallels between the two, both mechanistically and functionally. Moreover, an extensive literature, mainly focusing on the rodent hippocampus, demonstrates that impairment of LTP through genetic or pharmacological manipulations also impairs memory formation while modest enhancement of LTP through similar manipulations also enhances memory [37-40].

Some of the first groundbreaking cellular and molecular studies of synaptic plasticity and learning were conducted using *Aplysia* and have provided a significant conceptual platform for research on synaptic plasticity in the mammalian brain. A main idea derived from the work on *Aplysia* is that activity-dependent increases in synaptic strength may have different temporal phases that are characterized by distinct requirements for signal transduction, gene expression and protein synthesis [37]. This has led to the elucidation of at least two distinct temporal phases of LTP in the hippocampus: an "early" phase (E-LTP) and a longer lasting "late" phase (L-LTP). In an interesting parallel to STM formation, E-LTP is typically induced by a single high-frequency tetanus train of 100 Hz stimulation, lasts for approximately 1-3 hours and does not require protein synthesis. On the other hand, L-LTP is typically induced by 3 or 4 trains of 100 Hz stimulation, can exist for longer than 8 hours in hippocampal slices (41-43) and requires activation of transcription and protein synthesis [37, 41, 43-45], similar to LTM formation.

Studies using *Aplysia* [46-48] and *Drosophila* [49, 50] have demonstrated that members of the CREB family of transcription factors are necessary for long-lasting forms of synaptic plasticity and for LTM. In the mammalian CNS, an extensive number of reports have also supported a role for CREB in mediating long-term plasticity, although the situation may be somewhat complex as some forms of hippocampal LTP appeared to be intact in certain strains of CREB mutant mice (discussed later).

Activation of CREB During LTP

Indirect evidence for the involvement of CREB in LTP has come from the observation that induction of L-LTP, but not E-LTP, causes CRE-mediated transcription in slices from the hippocampus [51-53] and amygdala [54, 55] as well as in dissociated hippocampal neurons [13, 56] and the hippocampus *in vivo* [57, 58]. Interestingly, inhibition of cellular or nuclear PKA activity in neurons of the CA1 of the hippocampus blocks L-LTP as well as CRE-mediated gene expression [59]. Phosphorylation of CREB at Ser133 has been used as an indicator of CREB activity and has been shown to occur with either high- or low-frequency tetanic stimulation of cultured rodent hippocampal neurons in a NMDA receptor and calcium-dependent manner, highlighting the ability of multiple pathways, not just through PKA, to activate CREB-mediated transcription. NMDA receptor-dependent increases in phosphorylated CREB, but not total CREB, subsequent to the induction of LTP, has also been noted in hippocampal neurons from acute and organotypic slice preparations [60, 61]. Outside of the hippocampus, phosphorylation of CREB occurs within 2 hours of LTP induction in hippocampal-prefrontal cortical synapses *in vivo* [62].

LTP in CREB Mutant Mice

The direct involvement of CREB in mammalian LTP was first studied in animal models where the function of one or more CREB isoforms was eliminated (detailed descriptions of these animal models will be described in the next section). Using hippocampal slices from mutant mice lacking the α and δ isoforms of CREB (CREB$^{\alpha\delta}$ mice), Bourtchuladze *et al.* (1994) examined LTP by stimulating Schaffer collaterals with a train of 100 pulses at 100 Hz. Although the post-tetanic potential (measured 15-20 seconds after the tetanus) did not differ in CREB$^{\alpha\delta}$ mice, the resulting LTP was unstable. However, analagous to results from memory experiments that will be discussed in the next section, increasing the interval between tetani can overcome the LTP deficits observed in the CREB$^{\alpha\delta}$ mutants (unpublished observations). The temporal spacing between successive tetanic trains also alters the PKA-dependence of LTP [63, 64].

Similar to what was observed in CREB$^{\alpha\delta}$ mice, LTP produced by a spaced training protocol (3 stimulus trains of 100 Hz with a 10-min inter-train interval) was normal in the hippocampus of CREB^{comp} mice (CREB$^{\alpha\delta}$ mice that have only one allele for the CREB β isoform), as well as mice expressing a dominant-negative form of CREB in the CA1 region of the dorsal hippocampus (dCA1-KCREB mice) and two conditional CREB knockout mice (CREB^{CaMKCre7} and CREB^{NesCre}) [65-67]. Notably, however, the dCA1-KCREB mice showed deficient forskolin-induced, DA-regulated, and heterosynaptically-induced forms of hippocampal LTP [67, 68] as well as deficient LTP at corticostriatal synapses as reported by Pittenger *et al.* 2006. These findings highlight the importance of using several LTP-inducing paradigms to thoroughly investigate the role of CREB in synaptic plasticity. Furthermore, in light of observations that levels of transcriptional activator and repressor forms of CREM were up-regulated in CREB$^{\alpha\delta}$, CREB^{comp}, CREB^{CaMKCre7} and CREB^{NesCre} animals [66, 69, 70], the full genetic background of animals must also be taken into account.

LTP and CREB Overexpression

While decreasing CREB function disrupts some forms of LTP, increasing CREB function by overexpressing CREB acutely or in transgenic mice has been convincingly shown to enhance synaptic plasticity. In hippocampal slices from transgenic mice overexpressing a dominant active form of CREB (CREB-VP16), stimulation that normally induces E-LTP produced strong L-LTP [71]. Moreover, crossing these mice with increased CREB function with mice with a mutation in CBP rescues the deficit in L-LTP observed in the CBP mutant mice [72]. Similarly, transgenic mice overexpressing type-1 adenylyl cyclase (with elevated levels of pCREB) also showed L-LTP following a single train of high-frequency stimulation that produces only E-LTP in WT littermate controls [73]. Enhanced hippocampal LTP was also demonstrated in experiments where CREB-VP16 was acutely overexpressed in the hippocampus using a modified Sindbis virus [74].

CREB and LTD

In combination with LTP, LTD is thought to regulate the fine-tuning of synaptic weights that may be critical for learning [75, 76]. Bidirectional control of synaptic strength maintains synaptic efficacy within a dynamic operating range so that synapses can remain modifiable in response to electrical activity. LTD is typically induced by low-frequency stimulation [77-79] and, similar to LTP, is dependent on increases in intracellular calcium [80], is blocked by protein synthesis inhibitors [81] and is associated with increased levels of CREB phosphorylation in the hippocampus [56]. Furthermore, a dominant negative form of CREB inhibited LTD in corticostriatal synapses [82] as well as dissociated cerebellar neurons [81]. Differences in hippocampal LTD were also noted between CREB mutant mice (CREB^{comp}, CREB^{CaMKCre7}, CREB^{NesCre}) and wild type controls [65]. Mutants were reported to be more susceptible to low-frequency stimulation [65], suggesting important differences from wild type mice in the fine-tuning of synaptic plasticity. It should also be noted that in one report, acute over-expression of constitutively active CREB in the hippocampus enhanced LTP but not LTD [74].

Role of CREB in the Induction of Synaptic Plasticity

The mechanism by which CREB activity may contribute to synaptic plasticity has still to be determined in detail. The complement of genes which are induced by CREB during LTP/LTD are likely diverse and include *arc*, c-fos and brain derived neurotrophic factor (BDNF) [83-85]. The action of BDNF is an important factor in the maintenance of LTP [86]. In terms of functional changes consistent with some of the requirements for the induction of synaptic plasticity, CREB has been shown to increase the excitability of neurons, likely by increasing sodium currents and decreasing potassium currents [87, 88]. Additionally, increasing CREB activity can trigger a number of events that are critical for the ability for LTP to be induced, including an increase in NMDA receptor-mediated synaptic currents, the number of surface NMDA receptors and the ratio of NMDA/AMPA receptor-mediated currents [74, 89]. Collectively, these events are important for the formation of "silent" synapses, which are thought to be sites that are primed for the induction of LTP.

CREB AND MEMORY

Studies using a variety of animal models, both invertebrate and mammalian, have made fundamental strides towards resolving the role of CREB in LTM. Mammalian memory in rodent models are traditionally assessed using several memory tasks that differentially rely on distinct brain regions and include: cued and contextual fear conditioning, recognition memory and spatial memory. Analysis of the role for CREB in different memory tasks has involved correlative assessments (*e.g.* changes in CREB levels of phosphorylation state or CREB targets during and after memory acquisition), use of mutant mice (knockouts and transgenics overexpressing wild type or mutant forms of CREB) and the use of antisense and viral vectors to attenuate or enhance CREB function. Each approach alters CREB function in a distinct manner and, collectively, they provide a considerable body of evidence for the role of CREB-dependent transcription in LTM.

Manipulations of CREB for Assessment of LTM *In vivo*

Deletions of CREB by Homologous Recombination

The homozygous deletion of the three major CREB isoforms (α, β and δ; CREB^{null}) produced by a deletion in the bZIP domain in exon 10 of the *CREB* gene results in fully formed mice that die at birth [90]. However, inserting a neomycin-resistance cassette (neo) into exon 2 of the CREB gene produces viable mice that do not express the two main isoforms of CREB (α and δ) [91] leaving intact the translation of CREBβ which begins at exon 4. In fact, these CREB$^{\alpha\delta}$ mice have upregulated levels of CREBβ as well as CREM activator (τ) and repressor (α and β) isoforms [69]. Despite these upregulations, however, levels of CREB protein in the brain are reduced by 85-90% when compared to controls [92, 93] and CRE-DNA binding is virtually abolished [92, 94].

To minimize potential functional compensation of upregulated CREB β in LTM, an additional line of transgenic mice was created where a CREB$^{\alpha\delta}$ mouse was crossed with a mouse heterozygous for the CREB^{null} mutation, resulting in mice that are deficient for CREB α and δ and which carry only a single allele for the CREBβ isoform [66]. Similar to the CREB$^{\alpha\delta}$ mice, this strain of mice, referred to as CREB^{comp}, mature normally into adulthood and have upregulated levels of CREM.

Conditional CREB knockout mice have also been generated by crossing transgenic mice containing loxP recognition sequences around exon 10 of the *CREB* gene with transgenic mice expressing Cre recombinase driven under promoters for either αCaMKII or nestin [65, 70]. Recombination of the "floxed" CREB allele occurs in cells that produce sufficient levels of Cre recombinase, resulting in a gene encoding an unstable truncated CREB protein devoid of DNA-binding and dimerization domains. The αCaMKII promoter directs expression of Cre recombinase in postnatal excitatory forebrain neurons whereas the nestin promoter directs expression in all brain cells earlier in development (roughly embryonic day 12.5) [70]. Crossing the CREB^{loxP} mice with a line of transgenic mice expressing Cre recombinase fused to a C-terminal fragment of a progesterone receptor under the αCaMKII promoter produced mice (CREB^{CaMKCre7}) with a loss of CREB immunoreactivity in roughly 70-80% of CA1 (and other forebrain) neurons (65, 70), while CREB^{NesCre} mice show a loss of CREB in all brain regions [65, 70]. Similar to CREB$^{\alpha\delta}$ mice, these conditional CREB knockout mice have upregulated levels of CREM [70].

Decreasing CREB Function by Transgenic Expression of a Dominant Interfering form of CREB

Another method of disrupting CREB function involves the transgenic expression of a dominant negative allele that blocks the function of all CREB isoforms as well as CREM and ATF1. For instance, Rammes *et al.* (2000) developed transgenic mice that express a phosphorylation-defective CREB transgene (mCREB) in which serine 133 is mutated to an alanine (S133A). This mutated form of CREB cannot be activated by phosphorylation at Ser133, but still binds and occupies CRE sites, thus repressing endogenous CREB function [24]. Expression of this CREB repressor was constitutively driven by the αCaMKII promoter [95] and showed expression in the hippocampus, amygdala and cortex. Whether this manipulation caused a decrease of CRE-mediated transcription was not reported.

Inducible and Region Specific Manipulations of CREB Function

Temporal control over the mCREB expression was achieved by fusing the CREB repressor with a ligand-binding domain (LBD) of a human estrogen receptor with a G521R mutation (LBD^{G521R}) in transgenic mice [96]. The activity of this mutated LBD is regulated not by estrogen but by the synthetic ligand, tamoxifen [97-99]. In the absence of tamoxifen, the LBD^{G521R}-CREB^{S133A} fusion protein is bound to heat shock proteins and is therefore inactive [98]. However, administration of tamoxifen activates this inducible CREB repressor (CREB^{IR}) fusion protein, allowing it to compete with endogenous CREB and disrupt CRE-mediated transcription. Region specific expression was achieved by using a αCaMKII promoter, thus restricting expression primarily to the hippocampus, amygdala and cortex [96]. Importantly, unlike the CREB "knock-out" mice (CREB$^{\alpha\delta}$, CREB^{comp}, CREB^{NesCre} mice), CREB^{IR} mice do not show upregulated levels of CREM, perhaps due to the acute nature of the CREB disruption.

Pittenger and colleagues [67] used another dominant negative form of CREB to inducibly repress CREB function in forebrain neurons. K-CREB, a mutant form of CREB that dimerizes with other CREB family members but does not bind DNA [100, 101], was expressed under the αCaMKII promoter in combination with the tetracycline transactivator (tTA/*tetO*) system, in which expression of the transgene is turned off in the presence of doxycycline. One line of mice (referred to as dCA1-KCREB transgenic mice) showed expression of the transgene in dorsal CA1 regions of the hippocampus and was shown to disrupt CRE-mediated transcription [67].

Mice have also been engineered that have enhanced CREB function. Fusing CREB with the transactivation domain of a herpes simplex virus–derived transcriptional activator VP16 produced a form of CREB that is 25-fold more active that wild-type CREB *in vitro* [71]. Transgenic mice expressing this constitutively active form of CREB under the control of the αCaMKII promoter in combination with the tetracycline transactivator system were developed. CRE-mediated transcription in this mouse is increased as evidenced by enhanced expression of several CRE-regulated genes, including BDNF, dynorphin and c-fos [71]. Interestingly, expression of the VP16-CREB transgene for 3-4 weeks induces spontaneous seizures in mice while more prolonged expression (4-10 weeks) induced neurodegeneration and death [88].

Antisense Knockdown of CREB Function

Antisense oligonucleotides have also been used to acutely disrupt CREB function in a temporally and spatially restricted manner in rodent brains [102]. Synthetic oligonucleotides that are the genetic complement (antisense) of the mRNA encoding a specific protein can be introduced into neurons [103] and base-pair to their cognate mRNAs, thereby blocking the synthesis of specific proteins [104]. Injection of CREB antisense acutely and significantly decreased endogenous CREB protein (from 39-97%) in the brain region of interest [105, 106].

Viral Vectors to Manipulate CREB Function

Spatial and temporal control of the CREB function may also be achieved through the use of viral vectors. This method exploits the natural ability of viruses, such as herpes simplex virus (HSV) or Sindbis virus, to insert DNA specifically into neurons [107]. Replication-defective HSV vectors expressing wild-type (HSV-CREB) or mutant forms of CREB (HSV-mCREB) have been engineered. Infusing HSV-CREB into the brains of rats and mice not only increases the level of CREB protein but also CRE-mediated transcription [108, 109], while HSV-mCREB decreases CRE-mediated transcription [108, 109]. In addition to studies of the role of CREB in memory, this approach has also been used to show that CREB is a key transcription factor in the neuronal adaptation to drugs of abuse [109] and ocular dominance plasticity [110].

Role of CREB in Memory Formation

Conditioned Fear Memory

Fear conditioning is a common method of assessing memory in rodents. Typically, mice or rats are placed in a conditioning chamber and a footshock is delivered. In the discrete cue version of this task, a tone that co-terminates with the footshock is added thereby creating a learned association between the footshock and the auditory stimulus. Memory is assessed as the percentage of time rodents spend freezing (defined as the cessation of all movements except respiration) when placed back in the same conditioning chamber (referred to as contextual fear conditioning) or when the tone is played in a novel chamber (referred to as discrete cued fear conditioning). The neural systems critical for mediating fear conditioning have been well characterized. The amygdala is crucial for both cued and contextual fear conditioning as lesioning or inhibiting protein synthesis in this structure disrupts LTM [111-114]. The hippocampus likely plays an important role in contextual fear conditioning, but its' role is not as straightforward (see below).

The first study to examine the role of CREB in mammalian memory used the CREB$^{\alpha\delta}$ mice [115]. CREB$^{\alpha\delta}$ mice showed normal conditioned freezing to both a tone and context previously paired with footshock when tested shortly (<1 hour) after training, indicating that STM was normal in these mice when compared to wild type animals. However, both contextual and cued fear conditioning were impaired in another group of mice tested 24 hours after training [115]. The deficit in LTM for fear conditioning has been replicated using both freezing [116-118] and fear-potentiated startle [119] to measure conditioned fear. It is also important to note that the LTM deficit may be overcome by additional spaced training trials, highlighting the importance of the temporal dynamics of the training protocol to the CREB phenotype [118]. Strikingly, the restoration of LTM with spaced training is in parallel with restoration of normal LTP in brain slices from these mutant mice by spaced tetanic stimulation, as mentioned in the earlier section.

In CREB^{comp} mice (with only one CREB β allele), LTM both contextual and cued fear were impaired to a greater extent than in CREB$^{\alpha\delta}$ mice [66]. This finding further supports the involvement of CREB in long-term fear memory in that mice with a greater disruption in CREB function (CREB^{comp} versus CREB$^{\alpha\delta}$ mice) show a greater impairment in LTM.

Transgenic mice overexpressing the dominant negative form of CREB (mCREB) have also been tested using fear conditioning. One of 3 lines of mice that constitutively express the CREB repressor in the amygdala and hippocampus showed impaired LTM for cued fear conditioning [95]. In the other two transgenic lines, the apparent lack of effect of mCREB expression on cued fear memory may be explained by the fact that the transgene is expressed constitutively, possibly leading to a compensatory up-regulation by other CREB family members as shown for other mouse models in which CREB expression has been modified.

To overcome this potential limitation, fear memory was assessed in CREB^{IR} transgenic mice, allowing precise temporal control over the expression of the CREB repressor in the adult and preventing compensatory upregulation of other CREB isoforms [96]. CREB function was acutely and reversibly disrupted a short time (6 hours) prior to training and mice were tested 24 hours later, with presumably normal CREB function. LTM, but not STM, for both cued and contextual fear conditioning was impaired in these transgenic mice [96] indicating that functional CREB within a critical time window was required for LTM. Moreover, as the disruptions of CREB are temporary, these data also provide compelling evidence that the LTM phenotype observed in mutant mice with a chronic disruption of CREB (CREB$^{\alpha\delta}$ or CREB^{comp} mice) cannot be solely attributed to developmental abnormalities.

In light of disrupted LTM in CREB^{IR} mice, it was surprising to note the finding of normal LTM for contextual fear conditioning in 3 other types of CREB mutant mice. Normal LTM for contextual fear conditioning was observed in dCA1-KCREB mice (expressing a CREB repressor in the dorsal CA1 region of the hippocampus [67]) and two lines of conditional CREB knockout mice (CREB^{CaMKCre7}, with a deletion of CREB in 70-80% of CA1 neurons of the hippocampus; and CREB^{NesCre}, with a deletion of CREB in the entire nervous system [65]). However, previous results have suggested that the hippocampus is not strictly required for contextual fear conditioning in that non-hippocampal strategies can be utilized by the animal to acquire contextual information. While post-training lesions of dorsal hippocampus dramatically impair contextual fear conditioning, pre-training lesions do not [120, 121], implying the use of compensatory elemental (rather than contextual, hippocampal-based) strategies. Importantly, the three lines of CREB mutant mice that have normal LTM for contextual fear have chronic disruptions in CREB that could be analogous to "pre-training lesions" of CREB. Thus, unlike strategies to acutely disrupt CREB function, CREB expression is chronically impaired in both the CREB^{CaMKCre7} and CREB^{NesCre} mice. Even in the inducible dCA1-KCREB mice, the transgene is turned "on" for a significant period of time before training (*e.g.*, the mice are not fed doxycycline). Taken together, these findings suggest that this method of contextual fear conditioning may not be sufficiently sensitive to show potential LTM deficits in mice with more chronic pre-training perturbations in hippocampal CREB function.

To address the potential limitation of contextual fear conditioning, a variation of this task that critically relies on the hippocampus was used to evaluate the role of CREB in contextual LTM. Unlike standard contextual conditioning, this context pre-exposure task temporally separates the acquisition of context information from its association with the footshock [116, 122, 123]. Animals that were pre-exposed to a training context the day before receiving a shock immediately following replacement in that context show high levels of freezing, whereas animals not pre-exposed to the context show low levels of freezing following this immediate shock [116, 122, 123]. This phenomenon is sometimes referred to as the immediate shock deficit. Infusion of anisomycin, an inhibitor of protein translation, into the dorsal hippocampus immediately following context pre-exposure results in low freezing levels following the shock on the following day, indicating that protein synthesis in the dorsal hippocampus is vital for the context pre-exposure rescue of the immediate shock deficit [124]. Perhaps similarly, CREB^{αδ} mice show low levels of freezing following an immediate shock despite being pre-exposed to the training context the day before. Freezing levels in the CREB^{αδ} mice were similar to those observed in wild type mice that were not pre-exposed to the training context [116]. Therefore, using this task that critically depends on protein synthesis in the hippocampus, CREB was shown to be important for LTM for contextual memory.

The studies above clearly demonstrate deficits in long-term fear memory with disruptions of CREB function. In animals with intact CREB function, behavioural training that produces LTM for conditioned fear also increased the levels of phosphorylated CREB and CRE-regulated genes (such as C/EBP and BDNF) [125-132]. An increase in CRE-mediated transcription in the CA1 region of the dorsal hippocampus and amygdala following fear training has similarly been shown in mice expressing a CRE-lacZ reporter gene [127, 133]. Conversely, circadian increases in cAMP and pCREB in the hippocampus of mice were shown to correlate with the strength of contextual fear memory formation [134]. Although these data linking CREB activation to behavioral stimuli that induce LTM are correlative, they converge with the findings from animals with attenuated CREB function.

Further support for the relevance of CREB in LTM for fear has come from studies in which CREB was acutely over-expressed in rodents. An HSV-derived virus containing the gene for CREBα was used to increase the levels of CREB protein in a subpopulation of neurons in the lateral amygdala (~ 20-30% of neurons) [135]. Rats were trained for cued fear conditioning using a massed training protocol that typically produces weak STM but no or weak LTM. Compared to animals injected with HSV containing only a reporter gene, rats infused with HSV-CREB in the amygdala two days prior to training show enhanced LTM, similar to levels produced by spaced training. The finding of enhanced cued fear memory following infusion of HSV-CREB into the amygdala was also shown in a subsequent study [136]. Additionally, HSV-mediated CREB overexpression in a region of the auditory thalamus that projects to the lateral amygdala similarly increased LTM for fear [137]. Recently, it was also shown that infusing a virus encoding a constitutuvely active form of CREB (CREBY134F) into the CA1 or DG regions of the hippocampus increases memory for context fear conditioning [138]. Thus, acutely increasing CREB levels and function can enhance memory, a finding consistent with results in *Drosophila* (139) and *Aplysia* [46].

Recognition Memory (Object and Social Recognition)

Novel object recognition is a memory task that relies on the natural exploratory behavior of mice. Typically, training consists of exposing mice to two identical objects for a short period of time (*e.g.*, 15 min). Memory for this task is shown by mice spending more time exploring a novel object rather than the familiar object used in training. Two of the transgenic lines with disrupted CREB function, dCA1-KCREB mice [67] and CREB[IR] mice [140], have a deficit in LTM for object recognition. A similar behavioural deficit, along with reduced CREB-mediated transcription in the hippocampus during memory consolidation, has been observed in transgenic mice with inactivating mutations in the CREB binding domain (KIX domain) of CBP [141].

Consistent with previous data using a variety of learning tasks, spaced training, rather than massed training produces maximal LTM for object recognition [142]. Furthermore, spaced but not massed object recognition training is associated with increased levels of phosphorylated CREB and CRE-mediated transcription in the hippocampus and cortex [142]. However, in transgenic mice with decreased activity of protein phosphatase 1 (PP1) (that dephosphorylates CREB at Ser133) massed training alone is sufficient to produce both robust LTM and an increase in CRE-mediated transcription [142]. A similar pattern of results was also observed in transgenic mice that overexpress type-1 adenylyl cylase [73]. These mice show elevated levels of pCREB and LTM for object recognition following a single training trial whereas the wild type controls do not.

Similar results have also been found using a social recognition task. Training in this memory test consists of exposing a mouse to a conspecific. Memory is inferred if the amount of time a mouse spends exploring this familiar conspecifics is less than the time spent exploring a novel conspecific. LTM, but not STM, for social recognition is impaired in both the CREB[αδ] and CREB[IR] mice [143]. Therefore, using two types of recognition tasks, several groups of researchers have shown that CREB is activated following spaced training whereas disrupting CREB impaired LTM and increasing CREB function enhanced LTM.

Food Preference Memory

Rodents develop a preference for foods recently smelled on the breath of other rodents [144-146]. Training that produces LTM for social transmission of food preference increases the level of pCREB in the hippocampus [147]. Consistent with this, CREB[αδ] mice show intact STM, but impaired LTM, in this task [118]. Using a longer "training time" (10 minutes) with which to interact with the breath of a demonstrator mouse, Gass and colleagues (1998) showed normal LTM in both CREB[αδ] and CREB[comp] mice in a C57Bl\6 x FVB/N background. These results show that the deficits in CREB mutant mice are dependent on the training conditions of the tasks, and perhaps, even on the genetic background of the animals tested. This and other results demonstrate that CREB is not the only transcription factor mediating LTM formation, and, that under certain circumstances, the loss of CREB can be compensated by other genes.

Olfactory Memory

Memory may also be assessed using an olfactory conditioning task in neonatal rats. In this task an odor is paired with an appetitive (such as a stroke of the back) or an aversive (footshock) stimulus. Both appetitive and aversive olfactory conditioning are associated with an increase in pCREB levels in the olfactory bulbs [148, 149]. Moreover, disrupting CREB function (*via* infusion of CREB antisense oligonucleotides or HSV-mCREB) produces a specific LTM deficit for olfactory conditioning [149, 150].

Conditioned Taste Aversion Memory

In the conditioned taste aversion paradigm, ingestion of a novel taste is paired with transient sickness (produced by injection of lithium chloride, LiCl). Memory for this association is demonstrated by the animal avoiding that taste on subsequent presentations [151]. Previous studies show that LTM for conditioned taste aversion depends on protein synthesis in the amygdala [152, 153] and insular cortex [154].

The results from many studies that use different techniques to disrupt CREB function (mutant mice and antisense) converge to show that CREB is important in LTM for conditioned taste aversion. Thus, CREB[αδ] [152], CREB[IR] [152] and CREB[NesCre] [65] mice show disrupted LTM for conditioned taste aversion. Acutely disrupting CREB function in the amygdala through the use of antisense similarly disrupts LTM but not STM for conditioned taste aversion [106]. Furthermore, training that produces conditioned taste aversion (pairing the novel taste with LiCl) also induced robust CREB phosphorylation in the lateral nucleus of the amygdala [155]. Similar increases are not observed if rats are exposed to the taste or LiCl alone, indicating that activation of CREB is related to associative learning. Unlike some other memory tasks, conditioned taste aversion places few performance demands on the subject. Therefore conclusions regarding the role of CREB in LTM may be drawn independent of potential effects on other behaviors, such as motor behaviour.

Spatial Memory

In the hidden platform version of the Morris water maze rodents learn to find a platform submerged in a pool of opaque water by using spatial cues in the experimental room [156]. Typically, spatial memory is assessed during a probe trial in which the platform is removed and the percentage of time the animals spends searching in the location where the platform was previously positioned (target quadrant) is measured. This form of spatial memory is sensitive to hippocampal lesions [156, 157].

In the first study to use the Morris water maze to study the role of CREB in spatial memory, $CREB^{\alpha\delta}$ mice were trained once a day for 15 days [115]. As training progressed, the time to reach the platform grew shorter and shorter in wild type mice (indicating learning) while the latencies of $CREB^{\alpha\delta}$ mice remained long. As predicted from these training latencies, wild type mice searched selectively in the target quadrant during the probe test while $CREB^{\alpha\delta}$ mice searched randomly [115]. This finding has recently been replicated [158]. Similar to other memory tests discussed above, the role of CREB in spatial memory is sensitive to the training parameters. Thus, increasing the time between trials to 10 minutes or 1 hour (rather than 1 minute) overcomes the spatial memory deficit in $CREB^{\alpha\delta}$ mice [66, 118], implying additional strategies available for learning, independent of CREB.

Spatial memory deficits have also been observed using dCA1-KCREB mice which express the dominant-negative form of CREB in the hippocampal CA1 region [67]. Notably, dCA1-KCREB transgenic mice that were fed doxycycline for 2 weeks prior to training, thereby turning the transgene off, had normal spatial memory. In addition, spatial memory disruptions were found following infusion of CREB antisense into the dorsal hippocampus [105] or specifically into the hippocampal CA3 region [159]. Consistent with this, reduced levels of pCREB in transgenic mice lacking NFκB, a transcription factor that regulates expression of the catalytic subunit of PKA [160], was correlated with a deficit in spatial memory, as was reduced pCREB in the hippocampus of adult mice after prenatal administration of morphine [161].

Balschun and colleagues (2003) evaluated the impact of various CREB mutations in a large analysis of spatial memory using the Morris water maze. The percentage of time spent in the target quadrant during the probe trial for wild type mice (n=38) was 35.78%, $CREB^{\alpha\delta}$ (n=10) 28.5%, $CREB^{comp}$ (n=13) 26.39%, $CREB^{NesCre}$ (n=14) 31.11% and $CREB^{CaMKCre7}$ (n=12) 40.87%. Although these authors interpret this pattern of results as indicating that CREB is not necessary for spatial memory, an alternative interpretation is that the spatial memory phenotype is determined by the degree of CREB disruption. In support of this, the $CREB^{comp}$ mice show poorer spatial memory than the $CREB^{\alpha\delta}$ mice (see also Gass *et al.* 1988) and the $CREB^{NesCre}$ mice (with a virtually complete deletion in CREB) showed poorer spatial memory than the $CREB^{CaMKCre7}$ mice (with a loss of CREB in 70-80% of CA1 neurons). This latter finding of normal spatial memory in the $CREB^{CaMKCre7}$ mice suggests, therefore, that the remaining 20-30% of CA1 neurons with normal levels of CREB is sufficient to support normal spatial memory. Together, these findings suggest that CREB function may be disrupted in a larger portion of hippocampal neurons by transgenic overexpression of a dominant negative form of CREB (as in dCA1-KCREB transgenic mice) or CREB antisense than the limited pattern of CREB deletion observed in the hippocampus of $CREB^{CaMKCre7}$ mice.

CREB AND MEMORY DISORDERS

A number of cognitive disorders in humans have been directly associated with molecular lesions in the CREB signaling pathway (see [162] for review), including Coffin-Lowry Syndrome, Rubinstein-Taybi Syndrome and age-related cognitive decline. In addition, disruption in CREB signaling has been implicated in memory deficits associated with a number of diseases, secondary to the primary pathophysiology of the disorder. These include Alzheimer's disease, Huntington's disease and neurofibromatosis, among others.

Coffin-Lowry Syndrome and Rubinstein-Taybi Syndrome

Both Coffin-Lowry Syndrome and Rubinstein-Taybi Syndrome (RTS) are congenital disorders in which the CREB signaling pathway is directly implicated and which are characterized by severe cognitive deficiencies. Coffin-Lowry Syndrome is an X-linked disorder caused by mutations in the gene encoding RSK2, a protein kinase that phosphorylates CREB at the key Ser133 residue [163]. The intelligence of patients with Coffin-Lowry Syndrome is directly correlated with the capacity of their mutated RSK-2 to phosphorylate CREB [164]. RTS is a disorder characterized by severe mental retardation and physical abnormalities and is caused by chromosomal rearrangements, microdeletions and point mutations in one copy of the gene encoding CREB binding protein (CBP) [165].

Several groups have created mouse models of RTS to characterize cellular and molecular changes with this disorder as well as to examine potential therapeutic targets. In each of these models, CREB activity was affected by CBP mutations/deletions. Heterozygous mutant mice expressing one truncated *cbp* allele (acting as a dominant- negative) that lacked histone acetyltransferase (HAT) were shown to have normal learning and STM, but attenuated CREB-mediated transcription and hippocampal L-LTP along with reduced LTM for passive avoidance learning, object recognition and spatial memory [166-168]. Forebrain-restricted expression of a CBP mutant that lacked HAT activity resulted in mice that had impaired spatial and recognition LTM, but normal LTM for fear [169]. By comparison, in mice with haplo-insufficiency of CBP [72], deficits were observed in hippocampal L-LTP as well as LTM for both contextual and cued fear, although spatial memory appeared to be intact.

It is of note that in these models of RTS the impairment of object recognition LTM and L-LTP could be ameliorated by administering phosphodiesterase 4 (PDE4) inhibitors (drugs that increase cAMP levels thereby potentiating CREB function) [72, 166] and deficits in L-LTP and LTM for contextual fear could be rescued by inhibitors of histone deacetylase. Together, these findings indicate that at least some of the memory impairments observed in mouse models of RTS may be caused by a disruption of CREB-dependent transcription. Moreover, they suggest that targeting CREB function may be a viable therapeutic strategy in the treatment of RTS.

CREB and Age-Related Memory Decline

Although aging does not always result in memory impairments, aged animals, including humans, tend to show age-related memory deficits [170]. Age-related memory decline is correlated with hippocampal dysfunction [171-174] and aged humans and rodents both perform poorly on different versions of the Morris water maze [175].

The exact nature of the underlying biological mechanisms of age-related memory decline remains unknown. There is considerable evidence to support the view that age-related memory deficits may be caused by molecular changes in the absence of major structural alterations such as neuronal loss [176]. For instance, aged rodents show disruptions of several key members of the CREB signalling pathway, including CREB itself [177], pCREB [171], CBP [178] and PKA [179]. Importantly, the poor performance of rodents in the Morris water maze was correlated with decreased levels of CREB and pCREB in the hippocampus [180, 181]. In addition, the rate of "forgetting" of a socially transmitted food preference was found to be more rapid in aged rats and also correlated with reduced learning-dependent increases in hippopcampal pCREB [182]. Similarly, contextual fear conditioning is impaired in aged rats which also displayed reduced increases in pCREB during training when compared to younger animals [183, 184]. These findings suggest that the memory deficit associated with aging may be partially caused by decreased CREB function. Increasing CREB function may reverse age-related memory deficits. Accordingly, age-related memory decline was reversed in transgenic mice in which PP1 function was inhibited [142]. The effects of PP1 inhibition include an increase in CREB function brought about by decreased de-phosphorylation of CREB at S133. However, because PP1 inhibition affects a broad number of proteins, it is unclear whether the reversal of age-related memory decline may be directly attributable to increased CREB function.

CREB and Alzheimer's Disease

Alzheimer's disease (AD) is a progressive neurodegenerative disorder characterized by deterioration of cognitive and memory processes. Although the precise cause of AD is not known, evidence suggests that production and accumulation of β-amyloid peptide is central to disease pathogenesis [185]. Impairment of memory precedes neuronal death in patients and deficits in memory and LTP have been observed in mouse models of AD [186, 187]. Notably, induction of β-amyloid accumulation has been shown to decrease phosphorylation of CREB *in vivo* as well as in cultured neurons and slice preparations [187-190]. The reduced levels of pCREB appear to be related to disruption of multiple signaling pathways by β-amyloid. These disruptions are in the form of inhibition of the activity of several kinases that can phosphorylate CREB (PKA, ERK, guyanlylate kinase) [187, 188, 190, 191] as well as increased calpain (calcium-activated cysteine protease)-mediated degradation of a set of proteins that include CREB [192].

There is strong evidence to indicate that enhancement of CREB action can ameliorate the memory deficits associated with AD. In the APP/PS1 mouse model of AD (mice with mutations in amyloid precursor protein, APP, and presenilin 1, PS1), inhibition of calpain with the cysteine protease inhibitor E64 restored both pCREB levels and LTP in the hippocampus along with contextual fear memory and spatial memory [192]. Similarly, increasing ubiquitin C-terminal hydrolase L1 (UchL1) in the APP/PS1 mouse model overcame deficits in LTP and contextual fear memory. This recovery was correlated with UchL1-mediated increase in PKA activity and CREB phosphorylation [193]. In rats treated with β-amyloid to induce AD-like symptoms, treatment with a naturally occurring compound, nobiletin, restored memory impairments possibly by restoration of normal pCREB levels that were initially decreased by β-amyloid [189]. Collectively, these studies are striking in their ability to restore CREB function and memory-related deficits in disease models and provide clear therapeutic potential for the treatment of AD.

Other Disorders

Dysregulation of CREB has also been implicated in the symptoms of number of other disorders including neurofibromatosis, Creutzfeldt-Jakob disease and Huntington's disease. The evidence for the involvement of CREB is not as strong as with disorders such as RTS and AD and so will not be discussed in great detail. Patients with neurofibromatosis frequently have deficits in learning and storing spatial memories [194] and transgenic mice heterozygous for a neurofibrin mutation were similarly shown to display spatial learning deficits. These mice also had attenuated LTP and hyperphosphorylated ERK and CREB in the hippocampus [195]. In hippocampal slices derived from these mice, treatment with an ERK inhibitor reduced PCREB levels back to normal and restored L-LTP.

In animal models of Creutzfeldt-Jakob disease, CREB levels and function, respectively, were observed to be lower [196] and CREB dysfunction has been suggested to play a role in certain neurodegenerative disorders such as Huntington's disease [70]. Further evidence is required to establish a role for CREB in these diseases but it is intriguing to speculate that, for aspects of each of them, CREB function may be a viable therapeutic target.

CONCLUSIONS

There is now overwhelming evidence that CREB signaling plays a fundamental role in the molecular events that underlie long-term storage of different forms of memory in the brain. The transcriptional products of CREB activation are critical for the induction and maintenance of long term synaptic changes that likely form the basis for memory storage. While not discussed here, a role of CREB in neurogenesis [197, 198] coupled with recent findings that the production of new neurons may be involved in the formation of spatial memories [199] indicates an additional mechanism contributing to memory formation that involves CREB.

The relationship between altered CREB signaling and memory dysfunction in several disorders, such as those discussed above, indicates that this pathway may be an attractive target for therapeutic interventions. The ability to improve memory deficits in animal models using a number of compounds that affect the CREB pathway bodes well for the future development of viable treatments. It is important, however, to also understand further the nature of

CREB signaling. For example: a variety of factors such as trial spacing, genetic background and gene dosage impact the relative importance of CREB to memory and LTP. Under certain conditions, therefore, deficits in CREB signaling and memory can be circumvented, although it could be argued that the overall cognitive state of the affected animal is still impaired. What conditions would be required, then, so that CREB signaling can be optimally affected in the context of potential therapies?

In a similar context, the molecular complexity of CREB signaling needs to be further elucidated to understand the impact of targeting this pathway. Roughly 300 different stimuli are known to elicit phosphorylation of CREB at Ser133 (see [200]). How the signaling pathways elicited by these stimuli are co-ordinated and regulated to converge upon CREB requires further study. Particular attention needs to be paid to temporal, spatial and tissue specific regulation. As many promoters have binding sites for multiple transcription factors along with CREB, it is also possible that cooperative actions of these factors are required under certain conditions.

In addition, the downstream effectors of CREB need to be identified. The list of putative CREB target genes is large. Genome-wide analyses of CREB binding motifs indicated 1349 sites in the mouse genome and 1663 sites in the human genome [201]. Also, a recent study has shown more than 860 binding sites occupied by CREB/pCREB in the CNS after induction of an electroconvulsive seizure [25]. Vast numbers of putative CREB-regulated genes, and the enormous functional diversity of these genes, will critically influence the development of target-specific treatments aimed at alleviating memory dysfunction by increasing CREB function.

ACKNOWLEDGEMENTS

SAJ is a Canada Research Chair (Canada) in Molecular and Cellular Cognition and received research funding from the Canadian Institutes of Health Research (CIHR) and the Hospital for Sick Children. AJR is the recipient of a Restracomp Fellowship from the Hospital for Sick Children.

REFERENCES

[1]   Frankland PW, Bontempi B. The organization of recent and remote memories. Nat Rev Neurosci 2005; 6(2): 119-30.
[2]   Maren S, Quirk GJ. Neuronal signalling of fear memory. Nat Rev Neurosci 2004; 5(11): 844-52.
[3]   Stork O, Welzl H. Memory formation and the regulation of gene expression. Cell Mol Life Sci 1999; 55(4): 575-92.
[4]   Bailey CH, Chen M. Time course of structural changes at identified sensory neuron synapses during long-term sensitization in Aplysia. J Neurosci 1989; 9(5): 1774-80.
[5]   Davis HP, Squire LR. Protein synthesis and memory: a review. Psychol Bull 1984; 96(3): 518-59.
[6]   Matthies H. In search of cellular mechanisms of memory. Prog Neurobiol 1989; 32(4): 277-349.
[7]   Foulkes NS, Sassone-Corsi P. More is better: activators and repressors from the same gene. Cell 1992; 68(3): 411-4.
[8]   Hoeffler JP, Meyer TE, Yun Y, Jameson JL, Habener JF. Cyclic AMP-responsive DNA-binding protein: structure based on a cloned placental cDNA. Science 1988; 242(4884): 1430-3.
[9]   Rehfuss RP, Walton KM, Loriaux MM, Goodman RH. The cAMP-regulated enhancer-binding protein ATF-1 activates transcription in response to cAMP-dependent protein kinase A. J Biol Chem 1991; 266(28): 18431-4.
[10]   Busch SJ, Sassone-Corsi P. Dimers, leucine zippers and DNA-binding domains. Trends Genet 1990; 6(2): 36-40.
[11]   Fimia GM, De Cesare D, Sassone-Corsi P. Mechanisms of activation by CREB and CREM: phosphorylation, CBP, and a novel coactivator, ACT. Cold Spring Harb Symp Quant Biol 1998; 63: 631-42.
[12]   Bacskai BJ, Hochner B, Mahaut-Smith M, et al. Spatially resolved dynamics of cAMP and protein kinase A subunits in Aplysia sensory neurons. Science 1993; 260(5105): 222-6.
[13]   Bito H, Deisseroth K, Tsien RW. CREB phosphorylation and dephosphorylation: a Ca(2+)- and stimulus duration-dependent switch for hippocampal gene expression. Cell 1996; 87(7): 1203-14.
[14]   Chen RH, Sarnecki C, Blenis J. Nuclear localization and regulation of erk- and rsk-encoded protein kinases. Mol Cell Biol 1992; 12(3): 915-27.
[15]   Chow FA, Anderson KA, Noeldner PK, Means AR. The autonomous activity of calcium/calmodulin-dependent protein kinase IV is required for its role in transcription. J Biol Chem 2005; 280(21): 20530-8.
[16]   Dash PK, Karl KA, Colicos MA, Prywes R, Kandel ER. cAMP response element-binding protein is activated by Ca2+/calmodulin- as well as cAMP-dependent protein kinase. Proc Natl Acad Sci USA 1991; 88(11): 5061-5.
[17]   Finkbeiner S, Tavazoie SF, Maloratsky A, Jacobs KM, Harris KM, Greenberg ME. CREB: a major mediator of neuronal neurotrophin responses. Neuron 1997; 19(5): 1031-47.
[18]   Hagiwara M, Alberts A, Brindle P, et al. Transcriptional attenuation following cAMP induction requires PP-1-mediated dephosphorylation of CREB. Cell 1992; 70(1): 105-13.
[19]   Hardingham GE, Arnold FJ, Bading H. Nuclear calcium signaling controls CREB-mediated gene expression triggered by synaptic activity. Nat Neurosci 2001; 4(3): 261-7.

[20]   Sindreu CB, Scheiner ZS, Storm DR. Ca^{2+}-stimulated adenylyl cyclases regulate ERK-dependent activation of MSK1 during fear conditioning. Neuron 2007; 53(1): 79-89.

[21]   Takemoto-Kimura S, Terai H, Takamoto M, *et al.* Molecular cloning and characterization of CLICK-III/CaMKIgamma, a novel membrane-anchored neuronal Ca2+/calmodulin-dependent protein kinase (CaMK). J Biol Chem 2003; 278(20): 18597-605.

[22]   Wu GY, Deisseroth K, Tsien RW. Activity-dependent CREB phosphorylation: convergence of a fast, sensitive calmodulin kinase pathway and a slow, less sensitive mitogen-activated protein kinase pathway. Proc Natl Acad Sci USA 2001; 98(5): 2808-13.

[23]   Xing J, Ginty DD, Greenberg ME. Coupling of the RAS-MAPK pathway to gene activation by RSK2, a growth factor-regulated CREB kinase. Science 1996; 273(5277): 959-63.

[24]   Gonzalez GA, Montminy MR. Cyclic AMP stimulates somatostatin gene transcription by phosphorylation of CREB at serine 133. Cell 1989; 59(4): 675-80.

[25]   Tanis KQ, Duman RS, Newton SS. CREB binding and activity in brain: regional specificity and induction by electroconvulsive seizure. Biol Psychiatry 2008; 63(7): 710-20.

[26]   Zhang X, Odom DT, Koo SH, *et al.* Genome-wide analysis of cAMP-response element binding protein occupancy, phosphorylation, and target gene activation in human tissues. Proc Natl Acad Sci USA 2005; 102(12): 4459-64.

[27]   Cha-Molstad H, Keller DM, Yochum GS, Impey S, Goodman RH. Cell-type-specific binding of the transcription factor CREB to the cAMP-response element. Proc Natl Acad Sci USA 2004; 101(37): 13572-7.

[28]   Mouravlev A, Young D, During MJ. Phosphorylation-dependent degradation of transgenic CREB protein initiated by heterodimerization. Brain Res 2007; 1130(1): 31-7.

[29]   Wu H, Zhou Y, Xiong ZQ. Transducer of regulated CREB and late phase long-term synaptic potentiation. FEBS J 2007; 274(13): 3218-23.

[30]   Kovacs KA, Steullet P, Steinmann M, Do KQ, Magistretti PJ, Halfon O, *et al.* TORC1 is a calcium- and cAMP-sensitive coincidence detector involved in hippocampal long-term synaptic plasticity. Proc Natl Acad Sci USA 2007; 104(11): 4700-5.

[31]   Zhou Y, Wu H, Li S, *et al.* Requirement of TORC1 for late-phase long-term potentiation in the hippocampus. PLoS ONE 2006; 1: e16.

[32]   Xu W, Kasper LH, Lerach S, Jeevan T, Brindle PK. Individual CREB-target genes dictate usage of distinct cAMP-responsive coactivation mechanisms. EMBO J 2007; 26(12): 2890-903.

[33]   Wadzinski BE, Wheat WH, Jaspers S, *et al.* Nuclear protein phosphatase 2A dephosphorylates protein kinase A-phosphorylated CREB and regulates CREB transcriptional stimulation. Mol Cell Biol 1993; 13(5): 2822-34.

[34]   Impey S, Fong AL, Wang Y, *et al.* Phosphorylation of CBP mediates transcriptional activation by neural activity and CaM kinase IV. Neuron 2002; 34(2): 235-44.

[35]   Janknecht R, Nordheim A. MAP kinase-dependent transcriptional coactivation by Elk-1 and its cofactor CBP. Biochem Biophys Res Commun 1996; 228(3): 831-7.

[36]   Ravnskjaer K, Kester H, Liu Y, *et al.* Cooperative interactions between CBP and TORC2 confer selectivity to CREB target gene expression. EMBO J 2007; 26(12): 2880-9.

[37]   Kandel ER. The molecular biology of memory storage: a dialogue between genes and synapses. Science 2001; 294(5544): 1030-8.

[38]   Kaplan MP, Abel T. Genetic approaches to the study of synaptic plasticity and memory storage. CNS Spectr 2003; 8(8): 597-610.

[39]   Matynia A, Kushner SA, Silva AJ. Genetic approaches to molecular and cellular cognition: a focus on LTP and learning and memory. Annu Rev Genet 2002; 36: 687-720.

[40]   Silva AJ. Molecular and cellular cognitive studies of the role of synaptic plasticity in memory. J Neurobiol 2003; 54(1): 224-37.

[41]   Frey U, Huang YY, Kandel ER. Effects of cAMP simulate a late stage of LTP in hippocampal CA1 neurons. Science 1993; 260 (5114): 1661-4.

[42]   Huang YY, Li XC, Kandel ER. cAMP contributes to mossy fiber LTP by initiating both a covalently mediated early phase and macromolecular synthesis-dependent late phase. Cell 1994; 79(1): 69-79.

[43]   Nguyen PV, Abel T, Kandel ER. Requirement of a critical period of transcription for induction of a late phase of LTP. Science 1994; 265(5175): 1104-7.

[44]   Frey U, Krug M, Reymann KG, Matthies H. Anisomycin, an inhibitor of protein synthesis, blocks late phases of LTP phenomena in the hippocampal CA1 region in vitro. Brain Res 1988; 452(1-2): 57-65.

[45]   Stanton PK, Sarvey JM. Blockade of long-term potentiation in rat hippocampal CA1 region by inhibitors of protein synthesis. J Neurosci 1984; 4(12): 3080-8.

[46]   Bartsch D, Ghirardi M, Skehel PA, *et al. Aplysia* CREB2 represses long-term facilitation: relief of repression converts transient facilitation into long-term functional and structural change. Cell 1995; 83(6): 979-92.

[47]   Dash PK, Hochner B, Kandel ER. Injection of the cAMP-responsive element into the nucleus of *Aplysia* sensory neurons blocks long-term facilitation. Nature 1990; 345(6277): 718-21.

[48]   Martin KC, Casadio A, Zhu H, *et al.* Synapse-specific, long-term facilitation of aplysia sensory to motor synapses: a function for local protein synthesis in memory storage. Cell 1997; 91(7): 927-38.

[49]   Tully T. Physiology of mutations affecting learning and memory in *Drosophila*--the missing link between gene product and behavior. Trends Neurosci 1991; 14(5): 163-4.

[50]   Yin JC, Wallach JS, Del Vecchio M, *et al.* Induction of a dominant negative CREB transgene specifically blocks long-term memory in *Drosophila*. Cell 1994; 79(1): 49-58.

[51]   Ahmed T, Frey S, Frey JU. Regulation of the phosphodiesterase PDE4B3-isotype during long-term potentiation in the area dentata *in vivo*. Neuroscience 2004; 124(4): 857-67.

[52]   Lu YF, Kandel ER, Hawkins RD. Nitric oxide signaling contributes to late-phase LTP and CREB phosphorylation in the hippocampus. J Neurosci 1999; 19(23): 10250-61.

[53]   Matthies H, Schulz S, Thiemann W, *et al.* Design of a multiple slice interface chamber and application for resolving the temporal pattern of CREB phosphorylation in hippocampal long-term potentiation. J Neurosci Methods 1997; 78(1-2): 173-9.

[54]   Huang YY, Martin KC, Kandel ER. Both protein kinase A and mitogen-activated protein kinase are required in the amygdala for the macromolecular synthesis-dependent late phase of long-term potentiation. J Neurosci 2000; 20(17): 6317-25.

[55]   Lin CH, Yeh SH, Lu KT, Leu TH, Chang WC, Gean PW. A role for the PI-3 kinase signaling pathway in fear conditioning and synaptic plasticity in the amygdala. Neuron 2001; 31(5): 841-51.

[56]   Deisseroth K, Bito H, Tsien RW. Signaling from synapse to nucleus: postsynaptic CREB phosphorylation during multiple forms of hippocampal synaptic plasticity. Neuron 1996; 16(1): 89-101.

[57]   Davis S, Vanhoutte P, Pages C, Caboche J, Laroche S. The MAPK/ERK cascade targets both Elk-1 and cAMP response element-binding protein to control long-term potentiation-dependent gene expression in the dentate gyrus in vivo. J Neurosci 2000; 20(12): 4563-72.

[58]   Schulz S, Siemer H, Krug M, Hollt V. Direct evidence for biphasic cAMP responsive element-binding protein phosphorylation during long-term potentiation in the rat dentate gyrus *in vivo*. J Neurosci 1999; 19(13): 5683-92.

[59]   Matsushita M, Tomizawa K, Moriwaki A, Li ST, Terada H, Matsui H. A high-efficiency protein transduction system demonstrating the role of PKA in long-lasting long-term potentiation. J Neurosci 2001; 21(16): 6000-7.

[60]   Ahmed T, Frey JU. Plasticity-specific phosphorylation of CaMKII, MAP-kinases and CREB during late-LTP in rat hippocampal slices *in vitro*. Neuropharmacology 2005; 49(4): 477-92.

[61]   Leutgeb JK, Frey JU, Behnisch T. Single cell analysis of activity-dependent cyclic AMP-responsive element-binding protein phosphorylation during long-lasting long-term potentiation in area CA1 of mature rat hippocampal-organotypic cultures. Neuroscience 2005; 131(3): 601-10.

[62]   Hotte M, Thuault S, Dineley KT, Hemmings HC, Jr., Nairn AC, Jay TM. Phosphorylation of CREB and DARPP-32 during late LTP at hippocampal to prefrontal cortex synapses in vivo. Synapse 2007; 61(1): 24-8.

[63]   Scharf MT, Woo NH, Lattal KM, Young JZ, Nguyen PV, Abel T. Protein synthesis is required for the enhancement of long-term potentiation and long-term memory by spaced training. J Neurophysiol 2002; 87(6): 2770-7.

[64]   Woo NH, Duffy SN, Abel T, Nguyen PV. Temporal spacing of synaptic stimulation critically modulates the dependence of LTP on cyclic AMP-dependent protein kinase. Hippocampus 2003; 13(2): 293-300.

[65]   Balschun D, Wolfer DP, Gass P, *et al*. Does cAMP response element-binding protein have a pivotal role in hippocampal synaptic plasticity and hippocampus-dependent memory? J Neurosci 2003; 23(15): 6304-14.

[66]   Gass P, Wolfer DP, Balschun D, *et al*. Deficits in memory tasks of mice with CREB mutations depend on gene dosage. Learn Mem 1998; 5(4-5): 274-88.

[67]   Pittenger C, Huang YY, Paletzki RF, *et al*. Reversible inhibition of CREB/ATF transcription factors in region CA1 of the dorsal hippocampus disrupts hippocampus-dependent spatial memory. Neuron 2002; 34(3): 447-62.

[68]   Huang YY, Pittenger C, Kandel ER. A form of long-lasting, learning-related synaptic plasticity in the hippocampus induced by heterosynaptic low-frequency pairing. Proc Natl Acad Sci USA 2004; 101(3): 859-64.

[69]   Blendy JA, Kaestner KH, Schmid W, Gass P, Schutz G. Targeting of the CREB gene leads to up-regulation of a novel CREB mRNA isoform. EMBO J 1996; 15(5): 1098-106.

[70]   Mantamadiotis T, Lemberger T, Bleckmann SC, *et al*. Disruption of CREB function in brain leads to neurodegeneration. Nat Genet 2002; 31(1): 47-54.

[71]   Barco A, Alarcon JM, Kandel ER. Expression of constitutively active CREB protein facilitates the late phase of long-term potentiation by enhancing synaptic capture. Cell 2002; 108(5): 689-703.

[72]   Alarcon JM, Malleret G, Touzani K, *et al*. Chromatin acetylation, memory, and LTP are impaired in CBP+/- mice: a model for the cognitive deficit in Rubinstein-Taybi syndrome and its amelioration. Neuron 2004; 42(6): 947-59.

[73]   Wang H, Ferguson GD, Pineda VV, Cundiff PE, Storm DR. Overexpression of type-1 adenylyl cyclase in mouse forebrain enhances recognition memory and LTP. Nat Neurosci 2004; 7(6): 635-42.

[74]   Marie H, Morishita W, Yu X, Calakos N, Malenka RC. Generation of silent synapses by acute in vivo expression of CaMKIV and CREB. Neuron 2005; 45(5): 741-52.

[75]   Bear MF, Malenka RC. Synaptic plasticity: LTP and LTD. Curr Opin Neurobiol 1994; 4(3): 389-99.

[76]   Linden DJ, Connor JA. Long-term synaptic depression. Annu Rev Neurosci 1995; 18: 319-57.

[77]   Barrionuevo G, Schottler F, Lynch G. The effects of repetitive low frequency stimulation on control and "potentiated" synaptic responses in the hippocampus. Life Sci 1980 15; 27(24): 2385-91.

[78]   Dudek SM, Bear MF. Homosynaptic long-term depression in area CA1 of hippocampus and effects of N-methyl-D-aspartate receptor blockade. Proc Natl Acad Sci USA 1992; 89(10): 4363-7.

[79]   Staubli U, Lynch G. Stable depression of potentiated synaptic responses in the hippocampus with 1-5 Hz stimulation. Brain Res 1990; 513(1): 113-8.

[80]   Lisman JE. Three Ca^{2+} levels affect plasticity differently: the LTP zone, the LTD zone and no man's land. J Physiol 2001; 532(Pt 2): 285.

[81]   Ahn S, Ginty DD, Linden DJ. A late phase of cerebellar long-term depression requires activation of CaMKIV and CREB. Neuron 1999; 23(3): 559-68.

[82]   Pittenger C, Fasano S, Mazzocchi-Jones D, Dunnett SB, Kandel ER, Brambilla R. Impaired bidirectional synaptic plasticity and procedural memory formation in striatum-specific cAMP response element-binding protein-deficient mice. J Neurosci 2006; 26(10): 2808-13.

[83]   Impey S, McCorkle SR, Cha-Molstad H, *et al*. Defining the CREB regulon: a genome-wide analysis of transcription factor regulatory regions. Cell 2004; 119(7): 1041-54.

[84]   Kawashima T, Okuno H, Nonaka M, *et al*. Synaptic activity-responsive element in the Arc/Arg3.1 promoter essential for synapse-to-nucleus signaling in activated neurons. Proc Natl Acad Sci USA 2009; 106(1): 316-21.

[85]   Tao X, Finkbeiner S, Arnold DB, Shaywitz AJ, Greenberg ME. Ca^{2+} influx regulates BDNF transcription by a CREB family transcription factor-dependent mechanism. Neuron 1998; 20(4): 709-26.

[86]   Barco A, Patterson S, Alarcon JM, *et al*. Gene expression profiling of facilitated L-LTP in VP16-CREB mice reveals that BDNF is critical for the maintenance of LTP and its synaptic capture. Neuron 2005; 48(1): 123-37.

[87]   Dong Y, Green T, Saal D, *et al*. CREB modulates excitability of nucleus accumbens neurons. Nat Neurosci 2006; 9(4): 475-7.

[88]   Lopez de Armentia M, Jancic D, Olivares R, Alarcon JM, Kandel ER, Barco A. cAMP response element-binding protein-mediated gene expression increases the intrinsic excitability of CA1 pyramidal neurons. J Neurosci 2007; 27(50): 13909-18.

[89]   Huang YH, Lin Y, Brown TE, *et al*. CREB modulates the functional output of nucleus accumbens neurons: a critical role of N-methyl-D-aspartate glutamate receptor (NMDAR) receptors. J Biol Chem 2008; 283(5): 2751-60.

[90]   Rudolph D, Tafuri A, Gass P, Hammerling GJ, Arnold B, Schutz G. Impaired fetal T cell development and perinatal lethality in mice lacking the cAMP response element binding protein. Proc Natl Acad Sci USA 1998; 95(8): 4481-6.

[91]   Hummler E, Cole TJ, Blendy JA, *et al*. Targeted mutation of the CREB gene: compensation within the CREB/ATF family of transcription factors. Proc Natl Acad Sci USA 1994; 91(12): 5647-51.

[92]   Walters CL, Blendy JA. Different requirements for cAMP response element binding protein in positive and negative reinforcing properties of drugs of abuse. J Neurosci 2001; 21(23): 9438-44.

[93]   Walters CL, Kuo YC, Blendy JA. Differential distribution of CREB in the mesolimbic dopamine reward pathway. J Neurochem 2003; 87(5): 1237-44.

[94]   Pandey SC, Mittal N, Silva AJ. Blockade of cyclic AMP-responsive element DNA binding in the brain of CREB delta/alpha mutant mice. Neuroreport 2000; 11(11): 2577-80.

[95]   Rammes G, Steckler T, Kresse A, Schutz G, Zieglgansberger W, Lutz B. Synaptic plasticity in the basolateral amygdala in transgenic mice expressing dominant-negative cAMP response element-binding protein (CREB) in forebrain. Eur J Neurosci 2000; 12(7): 2534-46.

[96]   Kida S, Josselyn SA, de Ortiz SP, et al. CREB required for the stability of new and reactivated fear memories. Nat Neurosci 2002; 5(4): 348-55.

[97]   Danielian PS, White R, Hoare SA, Fawell SE, Parker MG. Identification of residues in the estrogen receptor that confer differential sensitivity to estrogen and hydroxytamoxifen. Mol Endocrinol 1993; 7(2): 232-40.

[98]   Feil R, Brocard J, Mascrez B, LeMeur M, Metzger D, Chambon P. Ligand-activated site-specific recombination in mice. Proc Natl Acad Sci USA 1996; 93(20): 10887-90.

[99]   Logie C, Stewart AF. Ligand-regulated site-specific recombination. Proc Natl Acad Sci USA 1995; 92(13): 5940-4.

[100]   Walton KM, Rehfuss RP, Chrivia JC, Lochner JE, Goodman RH. A dominant repressor of cyclic adenosine 3',5'-monophosphate (cAMP)-regulated enhancer-binding protein activity inhibits the cAMP-mediated induction of the somatostatin promoter in vivo. Mol Endocrinol 1992; 6(4): 647-55.

[101]   Xie S, Price JE, Luca M, Jean D, Ronai Z, Bar-Eli M. Dominant-negative CREB inhibits tumor growth and metastasis of human melanoma cells. Oncogene 1997; 15(17): 2069-75.

[102]   Wahlestedt C. Antisense oligonucleotide strategies in neuropharmacology. Trends Pharmacol Sci 1994; 15(2): 42-6.

[103]   Ogawa S, Pfaff DW. Application of antisense DNA method for the study of molecular bases of brain function and behavior. Behav Genet 1996; 26(3): 279-92.

[104]   Ghosh MK, Cohen JS. Oligodeoxynucleotides as antisense inhibitors of gene expression. Prog Nucleic Acid Res Mol Biol 1992; 42: 79-126.

[105]   Guzowski JF, McGaugh JL. Antisense oligodeoxynucleotide-mediated disruption of hippocampal cAMP response element binding protein levels impairs consolidation of memory for water maze training. Proc Natl Acad Sci USA 1997; 94(6): 2693-8.

[106]   Lamprecht R, Hazvi S, Dudai Y. cAMP response element-binding protein in the amygdala is required for long- but not short-term conditioned taste aversion memory. J Neurosci 1997; 17(21): 8443-50.

[107]   Simonato M, Manservigi R, Marconi P, Glorioso J. Gene transfer into neurones for the molecular analysis of behaviour: focus on herpes simplex vectors. Trends Neurosci 2000; 23(5): 183-90.

[108]   Barrot M, Olivier JD, Perrotti LI, et al. CREB activity in the nucleus accumbens shell controls gating of behavioral responses to emotional stimuli. Proc Natl Acad Sci USA 2002; 99(17): 11435-40.

[109]   Carlezon WA, Jr., Thome J, Olson VG, et al. Regulation of cocaine reward by CREB. Science 1998; 282(5397): 2272-5.

[110]   Mower AF, Liao DS, Nestler EJ, Neve RL, Ramoa AS. cAMP/Ca^{2+} response element-binding protein function is essential for ocular dominance plasticity. J Neurosci 2002; 22(6): 2237-45.

[111]   Davis M. The role of the amygdala in fear and anxiety. Annu Rev Neurosci 1992; 15: 353-75.

[112]   Fanselow MS, Gale GD. The amygdala, fear, and memory. Ann N Y Acad Sci 2003; 985: 125-34.

[113]   LeDoux JE. Emotion circuits in the brain. Annu Rev Neurosci 2000; 23: 155-84.

[114]   Schafe GE, LeDoux JE. Memory consolidation of auditory pavlovian fear conditioning requires protein synthesis and protein kinase A in the amygdala. J Neurosci 2000; 20(18): RC96.

[115]   Bourtchuladze R, Frenguelli B, Blendy J, Cioffi D, Schutz G, Silva AJ. Deficient long-term memory in mice with a targeted mutation of the cAMP-responsive element-binding protein. Cell 1994; 79(1): 59-68.

[116]   Frankland PW, Josselyn SA, Anagnostaras SG, Kogan JH, Takahashi E, Silva AJ. Consolidation of CS and US representations in associative fear conditioning. Hippocampus 2004;14(5): 557-69.

[117]   Graves L, Dalvi A, Lucki I, Blendy JA, Abel T. Behavioral analysis of CREB alphadelta mutation on a B6/129 F1 hybrid background. Hippocampus 2002; 12(1): 18-26.

[118]   Kogan JH, Frankland PW, Blendy JA, et al. Spaced training induces normal long-term memory in CREB mutant mice. Curr Biol 1997; 7(1): 1-11.

[119]   Falls WA, Kogan JH, Silva AJ, Willott JF, Carlson S, Turner JG. Fear-potentiated startle, but not prepulse inhibition of startle, is impaired in CREBalphadelta-/- mutant mice. Behav Neurosci 2000; 114(5): 998-1004.

[120]   Frankland PW, Cestari V, Filipkowski RK, McDonald RJ, Silva AJ. The dorsal hippocampus is essential for context discrimination but not for contextual conditioning. Behav Neurosci 1998; 112(4): 863-74.

[121]   Maren S, Aharonov G, Fanselow MS. Neurotoxic lesions of the dorsal hippocampus and Pavlovian fear conditioning in rats. Behav Brain Res 1997; 88(2): 261-74.

[122]   Matus-Amat P, Higgins EA, Barrientos RM, Rudy JW. The role of the dorsal hippocampus in the acquisition and retrieval of context memory representations. J Neurosci 2004; 24(10): 2431-9.

[123]   Wiltgen BJ, Sanders MJ, Behne NS, Fanselow MS. Sex differences, context preexposure, and the immediate shock deficit in Pavlovian context conditioning with mice. Behav Neurosci 2001; 115(1): 26-32.

[124]   Barrientos RM, O'Reilly RC, Rudy JW. Memory for context is impaired by injecting anisomycin into dorsal hippocampus following context exploration. Behav Brain Res 2002; 134(1-2): 299-306.

[125]   Bevilaqua LR, Cammarota M, Paratcha G, de Stein ML, Izquierdo I, Medina JH. Experience-dependent increase in cAMP-responsive element binding protein in synaptic and nonsynaptic mitochondria of the rat hippocampus. Eur J Neurosci 1999; 11(10): 3753-6.

[126]   Cammarota M, Bevilaqua LR, Ardenghi P, et al. Learning-associated activation of nuclear MAPK, CREB and Elk-1, along with Fos production, in the rat hippocampus after a one-trial avoidance learning: abolition by NMDA receptor blockade. Brain Res Mol Brain Res 2000; 76(1): 36-46.

[127]   Impey S, Smith DM, Obrietan K, Donahue R, Wade C, Storm DR. Stimulation of cAMP response element (CRE)-mediated transcription during contextual learning. Nat Neurosci 1998; 1(7): 595-601.

[128]   Ou LC, Gean PW. Transcriptional regulation of brain-derived neurotrophic factor in the amygdala during consolidation of fear memory. Mol Pharmacol 2007; 72(2): 350-8.

[129]   Stanciu M, Radulovic J, Spiess J. Phosphorylated cAMP response element binding protein in the mouse brain after fear conditioning: relationship to Fos production. Brain Res Mol Brain Res 2001; 94(1-2): 15-24.

[130]   Taubenfeld SM, Wiig KA, Bear MF, Alberini CM. A molecular correlate of memory and amnesia in the hippocampus. Nat Neurosci 1999; 2(4): 309-10.

[131]   Taubenfeld SM, Wiig KA, Monti B, Dolan B, Pollonini G, Alberini CM. Fornix-dependent induction of hippocampal CCAAT enhancer-binding protein [beta] and [delta] Co-localizes with phosphorylated cAMP response element-binding protein and accompanies long-term memory consolidation. J Neurosci 2001; 21(1): 84-91.

[132] Viola H, Furman M, Izquierdo LA, *et al.* Phosphorylated cAMP response element-binding protein as a molecular marker of memory processing in rat hippocampus: effect of novelty. J Neurosci 2000; 20(23): RC112.

[133] Athos J, Impey S, Pineda VV, Chen X, Storm DR. Hippocampal CRE-mediated gene expression is required for contextual memory formation. Nat Neurosci 2002; 5(11): 1119-20.

[134] Eckel-Mahan KL, Phan T, Han S, *et al.* Circadian oscillation of hippocampal MAPK activity and cAMP: implications for memory persistence. Nat Neurosci 2008 11(9): 1074-82.

[135] Josselyn SA, Shi C, Carlezon WA, Jr., Neve RL, Nestler EJ, Davis M. Long-term memory is facilitated by cAMP response element-binding protein overexpression in the amygdala. J Neurosci 2001; 21(7): 2404-12.

[136] Wallace TL, Stellitano KE, Neve RL, Duman RS. Effects of cyclic adenosine monophosphate response element binding protein overexpression in the basolateral amygdala on behavioral models of depression and anxiety. Biol Psychiatry 2004; 56(3): 151-60.

[137] Han JH, Yiu AP, Cole CJ, Hsiang HL, Neve RL, Josselyn SA. Increasing CREB in the auditory thalamus enhances memory and generalization of auditory conditioned fear. Learn Mem 2008; 15(6): 443-53.

[138] Restivo L, Tafi E, Ammassari-Teule M, Marie H. Viral-mediated expression of a constitutively active form of CREB in hippocampal neurons increases memory. Hippocampus 2009; 19(3): 228-34.

[139] Yin JC, Del Vecchio M, Zhou H, Tully T. CREB as a memory modulator: induced expression of a dCREB2 activator isoform enhances long-term memory in *Drosophila*. Cell 1995; 81(1): 107-15.

[140] Bozon B, Kelly A, Josselyn SA, Silva AJ, Davis S, Laroche S. MAPK, CREB and zif268 are all required for the consolidation of recognition memory. Philos Trans R Soc Lond B Biol Sci 2003; 358(1432): 805-14.

[141] Wood MA, Attner MA, Oliveira AM, Brindle PK, Abel T. A transcription factor-binding domain of the coactivator CBP is essential for long-term memory and the expression of specific target genes. Learn Mem 2006; 13(5): 609-17.

[142] Genoux D, Haditsch U, Knobloch M, Michalon A, Storm D, Mansuy IM. Protein phosphatase 1 is a molecular constraint on learning and memory. Nature 2002; 418(6901): 970-5.

[143] Kogan JH, Frankland PW, Silva AJ. Long-term memory underlying hippocampus-dependent social recognition in mice. Hippocampus 2000; 10(1): 47-56.

[144] Bunsey M, Eichenbaum H. Selective damage to the hippocampal region blocks long-term retention of a natural and nonspatial stimulus-stimulus association. Hippocampus 1995; 5(6): 546-56.

[145] Galef BG, Jr., Mason JR, Preti G, Bean NJ. Carbon disulfide: a semiochemical mediating socially-induced diet choice in rats. Physiol Behav 1988; 42(2): 119-24.

[146] Galef BG, Jr., Wigmore SW, Kennett DJ. A failure to find socially mediated taste aversion learning in Norway rats (R. norvegicus). J Comp Psychol 1983; 97(4): 358-63.

[147] Countryman RA, Orlowski JD, Brightwell JJ, Oskowitz AZ, Colombo PJ. CREB phsophorylation and c-Fos expression in the hippocampus of rats during acquisition and recall of a socially transmitted food preference. Hippocampus 2005; 15(1): 56-7.

[148] McLean JH, Harley CW, Darby-King A, Yuan Q. pCREB in the neonate rat olfactory bulb is selectively and transiently increased by odor preference-conditioned training. Learn Mem 1999; 6(6): 608-18.

[149] Zhang JJ, Okutani F, Inoue S, Kaba H. Activation of the cyclic AMP response element-binding protein signaling pathway in the olfactory bulb is required for the acquisition of olfactory aversive learning in young rats. Neuroscience 2003; 117(3): 707-13.

[150] Yuan Q, Harley CW, Darby-King A, Neve RL, McLean JH. Early odor preference learning in the rat: bidirectional effects of cAMP response element-binding protein (CREB) and mutant CREB support a causal role for phosphorylated CREB. J Neurosci 2003; 23(11): 4760-5.

[151] Garcia J, Kimeldorf DJ, Koelling RA. Conditioned aversion to saccharin resulting from exposure to gamma radiation. Science 1955; 122(3160): 157-8.

[152] Josselyn SA, Kida S, Silva AJ. Inducible repression of CREB function disrupts amygdala-dependent memory. Neurobiol Learn Mem 2004; 82(2): 159-63.

[153] Yamamoto T, Fujimoto Y. Brain mechanisms of taste aversion learning in the rat. Brain Res Bull 1991; 27(3-4): 403-6.

[154] Rosenblum K, Meiri N, Dudai Y. Taste memory: the role of protein synthesis in gustatory cortex. Behav Neural Biol 1993; 59(1): 49-56.

[155] Swank MW. Phosphorylation of MAP kinase and CREB in mouse cortex and amygdala during taste aversion learning. Neuroreport 2000; 11(8): 1625-30.

[156] Morris RG, Garrud P, Rawlins JN, O'Keefe J. Place navigation impaired in rats with hippocampal lesions. Nature 1982; 297(5868): 681-3.

[157] Sutherland RJ, Kolb B, Whishaw IQ. Spatial mapping: definitive disruption by hippocampal or medial frontal cortical damage in the rat. Neurosci Lett 1982 ; 31(3): 271-6.

[158] Hebda-Bauer EK, Luo J, Watson SJ, Akil H. Female CREBalphadelta- deficient mice show earlier age-related cognitive deficits than males. Neuroscience 2007; 150(2): 260-72.

[159] Florian C, Mons N, Roullet P. CREB antisense oligodeoxynucleotide administration into the dorsal hippocampal CA3 region impairs long- but not short-term spatial memory in mice. Learn Mem 2006; 13(4): 465-72.

[160] Kaltschmidt B, Ndiaye D, Korte M, *et al.* NF-kappaB regulates spatial memory formation and synaptic plasticity through protein kinase A/CREB signaling. Mol Cell Biol 2006; 26(8): 2936-46.

[161] Yang SN, Huang LT, Wang CL, *et al.* Prenatal administration of morphine decreases CREBSerine-133 phosphorylation and synaptic plasticity range mediated by glutamatergic transmission in the hippocampal CA1 area of cognitive-deficient rat offspring. Hippocampus 2003; 13(8): 915-21.

[162] Weeber EJ, Sweatt JD. Molecular neurobiology of human cognition. Neuron 2002; 33(6): 845-8.

[163] Trivier E, De Cesare D, Jacquot S, *et al.* Mutations in the kinase Rsk-2 associated with Coffin-Lowry syndrome. Nature 1996; 384(6609): 567-70.

[164] Harum KH, Alemi L, Johnston MV. Cognitive impairment in Coffin-Lowry syndrome correlates with reduced RSK2 activation. Neurology 2001; 56(2): 207-14.

[165] Petrij F, Giles RH, Dauwerse HG, *et al.* Rubinstein-Taybi syndrome caused by mutations in the transcriptional co-activator CBP. Nature 1995; 376(6538): 348-51.

[166] Bourtchouladze R, Lidge R, Catapano R, *et al.* A mouse model of Rubinstein-Taybi syndrome: defective long-term memory is ameliorated by inhibitors of phosphodiesterase 4. Proc Natl Acad Sci USA 2003; 100(18): 10518-22.

[167] Oike Y, Hata A, Mamiya T, *et al.* Truncated CBP protein leads to classical Rubinstein-Taybi syndrome phenotypes in mice: implications for a dominant-negative mechanism. Hum Mol Genet 1999; 8(3): 387-96.

[168] Wood MA, Kaplan MP, Park A, *et al.* Transgenic mice expressing a truncated form of CREB-binding protein (CBP) exhibit deficits in hippocampal synaptic plasticity and memory storage. Learn Mem 2005; 12(2): 111-9.

[169]   Korzus E, Rosenfeld MG, Mayford M. CBP histone acetyltransferase activity is a critical component of memory consolidation. Neuron 2004; 42(6): 961-72.

[170]   Barnes CA, Nadel L, Honig WK. Spatial memory deficit in senescent rats. Can J Psychol 1980; 34(1): 29-39.

[171]   Foster TC, Sharrow KM, Masse JR, Norris CM, Kumar A. Calcineurin links Ca^{2+} dysregulation with brain aging. J Neurosci 2001; 21(11): 4066-73.

[172]   Rosenzweig ES, Barnes CA. Impact of aging on hippocampal function: plasticity, network dynamics, and cognition. Prog Neurobiol 2003; 69(3): 143-79.

[173]   Small SA, Chawla MK, Buonocore M, Rapp PR, Barnes CA. Imaging correlates of brain function in monkeys and rats isolates a hippocampal subregion differentially vulnerable to aging. Proc Natl Acad Sci USA 2004; 101(18): 7181-6.

[174]   Uttl B, Graf P. Episodic spatial memory in adulthood. Psychol Aging 1993; 8(2): 257-73.

[175]   Newman MC, Kaszniak AW. Spatial memory and aging" performance on a human analog Morris water maze. Aging, Neuropsychol Cogniton 2000; 7(2): 86-93.

[176]   Foster TC. Involvement of hippocampal synaptic plasticity in age-related memory decline. Brain Res Brain Res Rev 1999; 30(3): 236-49.

[177]   Asanuma M, Nishibayashi S, Iwata E, *et al*. Alterations of cAMP response element-binding activity in the aged rat brain in response to administration of rolipram, a cAMP-specific phosphodiesterase inhibitor. Brain Res Mol Brain Res 1996; 41(1-2): 210-5.

[178]   Chung YH, Kim EJ, Shin CM, *et al*. Age-related changes in CREB binding protein immunoreactivity in the cerebral cortex and hippocampus of rats. Brain Res 2002; 956(2): 312-8.

[179]   Karege F, Lambercy C, Schwald M, Steimer T, Cisse M. Differential changes of cAMP-dependent protein kinase activity and 3H-cAMP binding sites in rat hippocampus during maturation and aging. Neurosci Lett 2001; 315(1-2): 89-92.

[180]   Brightwell JJ, Gallagher M, Colombo PJ. Hippocampal CREB1 but not CREB2 is decreased in aged rats with spatial memory impairments. Neurobiol Learn Mem 2004; 81(1): 19-26.

[181]   Porte Y, Buhot M, Mons N. Alteration of CREB phosphorylation and spatial memory deficits in aged 129T2/Sv mice. Neurobiol Aging 2008; 29: 1533-46.

[182]   Countryman RA, Gold PE. Rapid forgetting of social transmission of food preferences in aged rats: relationship to hippocampal CREB activation. Learn Mem 2007; 14(5): 350-8.

[183]   Kudo K, Wati H, Qiao C, Arita J, Kanba S. Age-related disturbance of memory and CREB phosphorylation in CA1 area of hippocampus of rats. Brain Res 2005; 1054(1): 30-7.

[184]   Monti B, Berteotti C, Contestabile A. Dysregulation of memory-related proteins in the hippocampus of aged rats and their relation with cognitive impairment. Hippocampus 2005; 15(8): 1041-9.

[185]   Hardy J, Selkoe DJ. The amyloid hypothesis of Alzheimer's disease: progress and problems on the road to therapeutics. Science 2002; 297(5580): 353-6.

[186]   Sant'Angelo A, Trinchese F, Arancio O. Usefulness of behavioral and electrophysiological studies in transgenic models of Alzheimer's disease. Neurochem Res 2003; 28(7): 1009-15.

[187]   Vitolo OV, Sant'Angelo A, Costanzo V, Battaglia F, Arancio O, Shelanski M. Amyloid beta -peptide inhibition of the PKA/CREB pathway and long-term potentiation: reversibility by drugs that enhance cAMP signaling. Proc Natl Acad Sci USA 2002; 99(20): 13217-21.

[188]   Ma QL, Harris-White ME, Ubeda OJ, *et al*. Evidence of Abeta- and transgene-dependent defects in ERK-CREB signaling in Alzheimer's models. J Neurochem 2007; 103(4): 1594-607.

[189]   Matsuzaki K, Yamakuni T, Hashimoto M, *et al*. Nobiletin restoring beta-amyloid-impaired CREB phosphorylation rescues memory deterioration in Alzheimer's disease model rats. Neurosci Lett 2006; 400(3): 230-4.

[190]   Puzzo D, Vitolo O, Trinchese F, Jacob JP, Palmeri A, Arancio O. Amyloid-beta peptide inhibits activation of the nitric oxide/cGMP/cAMP-responsive element-binding protein pathway during hippocampal synaptic plasticity. J Neurosci 2005; 25(29): 6887-97.

[191]   Liang Z, Liu F, Grundke-Iqbal I, Iqbal K, Gong CX. Down-regulation of cAMP-dependent protein kinase by over-activated calpain in Alzheimer disease brain. J Neurochem 2007; 103(6): 2462-70.

[192]   Trinchese F, Fa M, Liu S, *et al*. Inhibition of calpains improves memory and synaptic transmission in a mouse model of Alzheimer disease. J Clin Invest 2008; 118(8):2796-807.

[193]   Gong B, Cao Z, Zheng P, *et al*. Ubiquitin hydrolase Uch-L1 rescues beta-amyloid-induced decreases in synaptic function and contextual memory. Cell 2006; 126(4): 775-88.

[194]   Hyman SL, Shores A, North KN. The nature and frequency of cognitive deficits in children with neurofibromatosis type 1. Neurology 2005; 65(7): 1037-44.

[195]   Guilding C, McNair K, Stone TW, Morris BJ. Restored plasticity in a mouse model of neurofibromatosis type 1 via inhibition of hyperactive ERK and CREB. Eur J Neurosci 2007; 25(1): 99-105.

[196]   Rodriguez A, Ferrer I. Expression of transcription factors CREB and c-Fos in the brains of terminal Creutzfeldt-Jakob disease cases. Neurosci Lett 2007; 421(1): 10-5.

[197]   Nakagawa S, Kim JE, Lee R, *et al*. Localization of phosphorylated cAMP response element-binding protein in immature neurons of adult hippocampus. J Neurosci 2002; 22(22): 9868-76.

[198]   Nakagawa S, Kim JE, Lee R, *et al*. Regulation of neurogenesis in adult mouse hippocampus by cAMP and the cAMP response element-binding protein. J Neurosci 2002; 22(9): 3673-82.

[199]   Kee N, Teixeira CM, Wang AH, Frankland PW. Preferential incorporation of adult-generated granule cells into spatial memory networks in the dentate gyrus. Nat Neurosci 2007; 10(3): 355-62.

[200]   Johannessen M, Delghandi MP, Moens U. What turns CREB on? Cell Signal 2004; 16(11): 1211-27.

[201]   Conkright MD, Guzman E, Flechner L, Su AI, Hogenesch JB, Montminy M. Genome-wide analysis of CREB target genes reveals a core promoter requirement for cAMP responsiveness. Mol Cell 2003; 11(4): 1101-8.

CHAPTER 2

The Role of CREB in Neuronal Plasticity, Learning and Memory, and in Neuropsychiatric Disorders

Ángela Fontán-Lozano, Rocío Romero-Granados, Eva M. Pérez-Villegas and Ángel M. Carrión[*]

Departamento de Fisiología, Anatomía y Biología Celular, Universidad Pablo de Olavide, Carretera de Utrera Km 1, 41013 Sevilla, Spain

Abstract: A characteristic of higher organisms is their ability to learn by experience in order to adapt their behaviour. Such plasticity is largely the result of the brain's ability to convert transient stimuli into long-lasting alterations in neuronal structure and function. This process is complex and involves changes in receptor trafficking, local mRNA translation, protein turnover and gene synthesis. Here, we will review how changes in neuronal activity trigger CREB-dependent gene expression in order to provoke more persistent changes in neuronal function. Interestingly, CREB activity is altered in some psychiatric disorder such as anxiety and depression, which suggest that this programme is essential for the correct functioning of the brain.

Keywords: Transcription factors, CREB, CBP, CREB target genes, Acetylation, Transduction signaling, Learning, Long-term memory, Consolidation, LTP, Neurodegeneration, Anxiety, Depression.

Environmental changes cause the modification of neurotransmitter release at specific synapses in the brain, affecting electrical and biochemical signalling events in the postsynaptic neuron. One of the most prominent of these signalling events involves a rapid and transient rise in calcium within the postsynaptic terminal. This local increase in calcium may produce a number of short-term and long-term synapse specific alterations, including the insertion or removal of receptor subunits from the membrane, modifications of synaptic protein function, and modifications in the protein content at the synapse (*via* the balance between translation and degradation). Together, such modifications lead to changes in synaptic function [reviewed in 1-3]. In addition, changes in electrical activity at the postsynaptic terminal initiate a cascade of intracellular signals that reach the nucleus, provoking the activation of new genetic programmes that affect more persistent events such as dendrite and synapse development, or neuronal plasticity [reviewed in 4]. The mechanisms underlying these activity-dependent transcriptional events and the nature of the gene expression programmes they induce have been the subject of intense investigation, leading to the identification of a number of activity-regulated transcription factors and effector genes. A series of observations in the past decade identified CREB-mediated transcription as a critical mediator of adaptive responses in the nervous system. Indeed, the potential role of the CREB (cAMP response element binding) protein in memory has been an active area of study. Moreover, recent studies into the role of CREB in central nervous system (CNS) physiology have begun to yield insights into the role of experience in CNS development and human disorders of cognition [5].

MOLECULAR BASIS OF LONG-LASTING ALTERATIONS IN CENTRAL NERVOUS SYSTEM

Protein Synthesis is Required for Long-Lasting Changes in the Nervous System

The theory of memory consolidation, the post-learning processes of memory stabilization, has guided contemporary research into the neurobiological basis of learning and memory. These changes are similar in some senses to those happening during long-term modifications in cell function, and *de novo* protein synthesis is one of the critical

*Address correspondence to Ángel M. Carrión: División de Neurociencias, Universidad Pablo de Olavide, Carretera de Utrera Km. 1, Sevilla-41013, Spain; Tel: +34-954-977503; Fax: +34-954-349375; E-mail: amancar@upo.es

processes related with the stabilization of such modifications. For this reason, protein synthesis inhibitors became an important tool in research into memory and consolidation. Indeed, by administering the potent protein synthesis inhibitor puromycin to gold fish, long-term memory was seen to require *de novo* protein synthesis in contrast to short-term memory [6]. Others studies [7, 8] strengthened the currently accepted model of memory consolidation, in which initially weak connections between newly recruited neurons are strengthened and become stable in a manner dependent on *de novo* protein synthesis. Indeed, the cellular responses necessary include the activation of second messenger systems, new RNA transcription and protein synthesis.

Immediate Early Genes in Long-Term Plasticity

Neurons propagate electrical impulses over distance (action potentials) and they communicate through synapses using chemical messengers (neurotransmitters). Complex signal transduction machinery converts these chemical signals into electric potentials and induces long-lasting modifications of certain cell properties. These mechanisms contribute to the unique features of neurons as elements in a signalling network or circuit. The activation of a neuron and its recruitment into an ensemble that represents a particular memory initiates mechanisms involved in the long-term modification of synaptic responses aimed at maintaining the coherence of that ensemble. Gene expression, and especially that of immediately early genes (IEGs), is one step in the cascade of cellular events underlying this mechanism [9-11]. IEGs are genes whose transcription can be induced in the presence of protein synthesis inhibitors and thus, their expression does not require *de novo* protein synthesis. The tight coupling of IEG expression to the patterning of synaptic activity and their putative cellular functions make them attractive candidates to influence the processes underlying long-term synaptic plasticity.

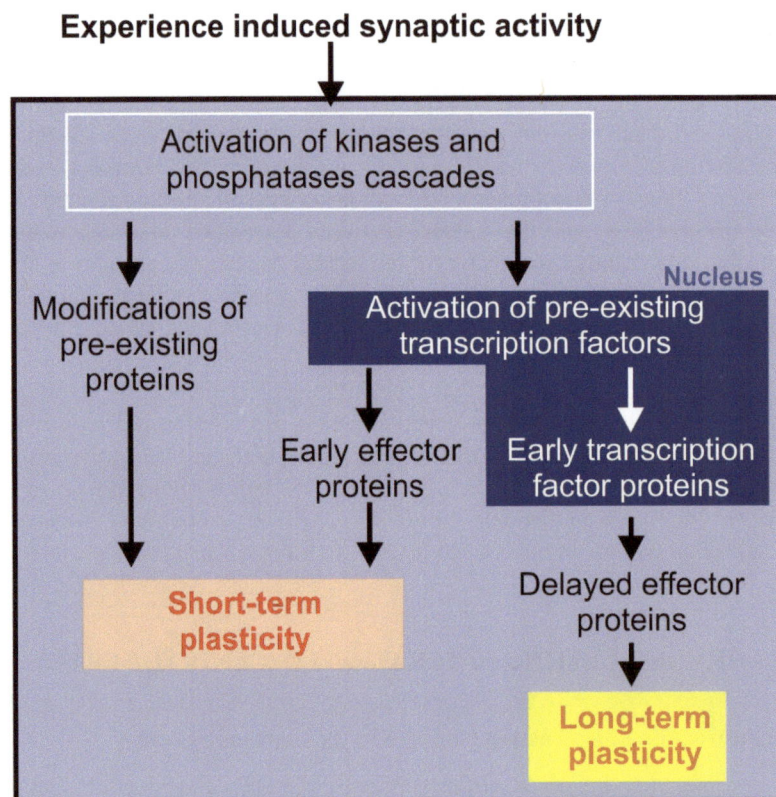

Fig. (1). Proposed model of experience-dependent gene expression in synaptic plasticity. Synaptic activity driven by experience affects the levels of intracellular second messengers, in turn activating cellular kinases and phosphatases. These enzymes modulate the activity of a wide range of cellular proteins, including synaptic components and nuclear transcription factors. In the nucleus, the activation of transcription factors initiates cascades of gene expression, some of which are required to establish long-term plasticity.

IEG transcription is initiated by patterned synaptic activity that induces long-term synaptic plasticity [12, 13]. Neurotransmitter binding to postsynaptic receptors, or action potential firing following the integration of postsynaptic potentials [14], provokes an influx of extracellular Ca^{2+} or a release of Ca^{2+} from intracellular stores [15] and activates signal transduction cascades that involve postsynaptic second-messengers and the activation of protein kinases and phosphatases (Fig. 1). Some of these kinases regulate nuclear gene expression [4, 16], including protein kinase A [17, 18], mitogen-activated protein kinase [19-22], and calcium and calmodulin dependent kinase [23-25]. These activated kinases target specific transcription factors in the nucleus, such as the cAMP response element binding protein (CREB) and serum response factor (SRF), and these transcription factors then bind to promoter regions of IEGs (CREB to CRE and SRF to SRE) confer responsiveness to second messengers or growth factors [18, 26]. In parallel, transcription factors such as NFAT and NFkB, which are localized in the cytoplasm before to stimulation, are activated by dephosphorylation and phosphorylation respectively, which in turn lead to the translocation of NFAT and NF-kB to the nucleus where they activate transcription [27, 28]. These activated transcription factors are then capable of recruiting the transcriptional machinery [24, 26, 29], thereby initiating IEG transcription. Indeed, IEG transcription has been observed within minutes of stimulation both *in vitro* [30] and *in vivo* [31, 32].

In the brain, studies on IEGs have focused on activity-induced transcription factors, such as c-fos, c-jun, and zif268 [reviewed in 11]. Subsequent cloning of brain/neuron-specific IEGs identified other proteins with a diverse range of cellular functions [33, 34]. These IEGs were called effector IEGs and they include Arc, Homer 1a, tissue-plasminogen activator (TPA), Narp and BDNF. By definition, IEGs that are transcription factors would be expected to drive the expression of delayed effector genes [9, 35], and such delayed effector genes may play specific roles in neurotransmission, cell maintenance and plasticity. It has also been speculated that regulatory IEGs play a role in metaplasticity or function as coincidence detectors [9, 10]. Effector IEGs have a wide range of cellular functions, including those related to cell growth (BDNF, Narp), intracellular signalling (RheB, RGS-2, Homer 1a), synaptic modification or other structural changes (Arc, Homer 1a, Narp, TPA, BDNF), metabolism (COX-2), or synaptic homeostasis (Arc, Homer 1a) [10, 33, 36]. These functions are compatible with the types of modifications thought to underlie synaptic plasticity. Long-term modifications of activity in the nervous system involve several known transcription factors. Hence, here we shall focus on the important role of CREB in synaptic plasticity related to information storage.

CREB FAMILY TRANSCRIPTION FACTORS: MOLECULAR STRUCTURE

CREB belongs to the bZIP superfamily of transcription factors characterized by the presence of at least one 7 amino acid leucine rich dimerization region called a leucine zipper. The CREB (or CREB/ATF) family of transcription factors includes three homologous genes (Fig. 2): *creb*, *crem* (cAMP response element modulator) and *atf-1* (activating transcription factor-1) [37]. While both CREBs and ATF-1 are ubiquitously expressed, CREMs proteins are mainly present in the neuroendocrine system. Like all bZIP transcription factors, CREB family members contain a C-terminal basic domain that mediates DNA binding, as well as the leucine zipper domain that facilitates dimerization. This structure enables proteins of the bZIP family to bind to their DNA binding sequences as dimers. In addition, the sequence of each bZIP domain also governs whether these proteins form homodimers or heterodimers.

There is a high degree of similarity between CREB, CREM and ATF-1, especially within the bZIP domain, consistent with the fact that these factors can form both homo- and heterodimers. Indeed, each protein can bind to the same *cis*-regulatory element [reviewed in 18, 38, 39]. The classical sequences known to be specifically recognized by CREB are palindromic octanucleotide sequences, TGACGTCA, referred to as cAMP response elements (CRE) [40, 41]. However, structural differences between different CREs significantly influence the DNA-protein interaction. Thus, particular CRE interact in a qualitatively different manner with different members of the CREB family, and while the DNA-binding stability of ATF1 and CREB differ dramatically (ATF1 binding is highly unstable), this is really dependent on the particular CRE in question [42]. Moreover, CREB phosphorylation affects its DNA-binding activity in a CRE-dependent manner [43, 44]. Indeed, it must be born in mind that there are two classes of CREs: symmetric sites that have high affinity for CREB and where DNA-binding is not stimulated by phosphorylation; and asymmetric sites with low affinity for unphosphorylated CREB and high affinity for phosphorylated CREB [44].

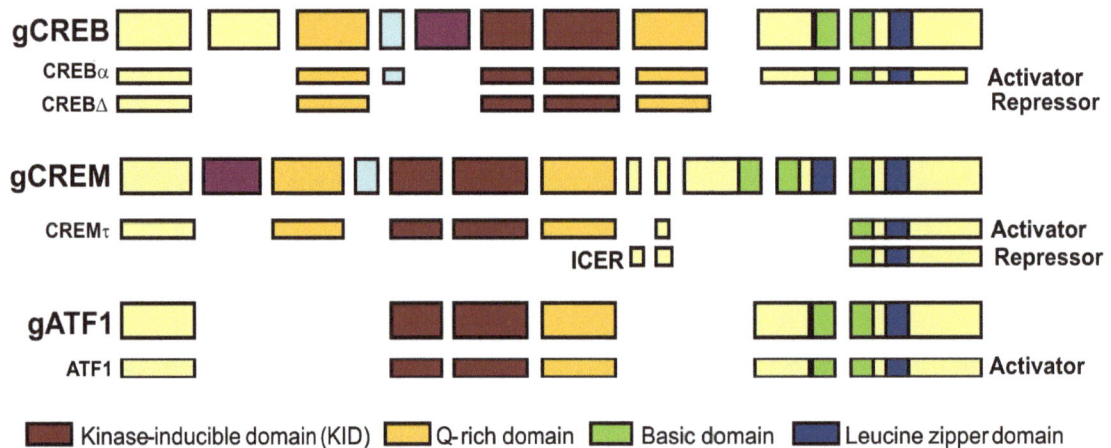

Fig. (2). Genomic organization of the mammalian CREB family. Consensus alignment of the genomic loci of the human cAMP response element (CRE)-binding protein (gCREB), the cAMP response element modulator (gCREM) and the activating transcription factor 1 (gATF1). The homologous domains are in colour (see key). Several alternatively spliced exons exist in the CREB and CREM genes, and some of the alternative splice products with divergent activating properties are shown.

While the bZIP domain mediates DNA binding and dimerization, the remaining domains of CREB family members serve to facilitate interactions with co-activators and the components of the transcriptional machinery that are ultimately required for RNA synthesis (the functional domains of CREB and its relatives [45-48] are represented in Fig. **2**). Moreover, alternative splicing of the CREB and CREM genes generates a vast repertoire of CREB family members, including several isoforms with distinct repressor or activator features [39]. The transactivation domain of CREB is bipartite, including two glutamine rich domains, referred to as Q1 and Q2/CAD (constitutive active domain), which are separated by the kinase inducible domain (KID). The Q2 domain mediates interactions with a component of the TFIID complex, whereas the KID seems to promote isomerisation by recruiting the CREB binding protein (CBP) co-activator and p300 to the promoter. While Q2/CAD interactions may be responsible for facilitating stimulus-independent CRE-driven gene expression [49-51], the KID region is only active when it is phosphorylated at Ser-133, which occurs in a stimulus-inducible manner and that is critical for the activation of CREB [39, 52]. CBP and p300 are transcriptional co-activators with intrinsic histone acetyltransferase activity, which is necessary for enhancing transcription through chromatin remodelling [53, 54]. The functional significance of these CREB domains and their respective interacting molecules will be considered in greater detail below.

CREB ACTIVATION BY EXTRACELLULAR SIGNALS DURING SYNAPTIC PLASTICITY

Since the identification of CREB as an inducible regulator of transcription, many studies have focused on the molecular mechanisms underlying CREB activation. Accordingly, many of the extracellular stimuli that activate CREB *via* signal transduction have been identified. In this way, Ser-133 of the KID domain has been identified as a key regulatory site for CREB to function as a stimulus-dependent transcriptional activator. In neurons, CREB phosphorylation occurs under a wide variety of circumstances, all of which are initiated by external stimuli. Indeed, under normal conditions CREB is activated in neurons by growth factors, depolarization and synaptic activity. In this section, we will consider the pathways that connect these external stimuli to CREB-activated gene expression (Fig. **3**).

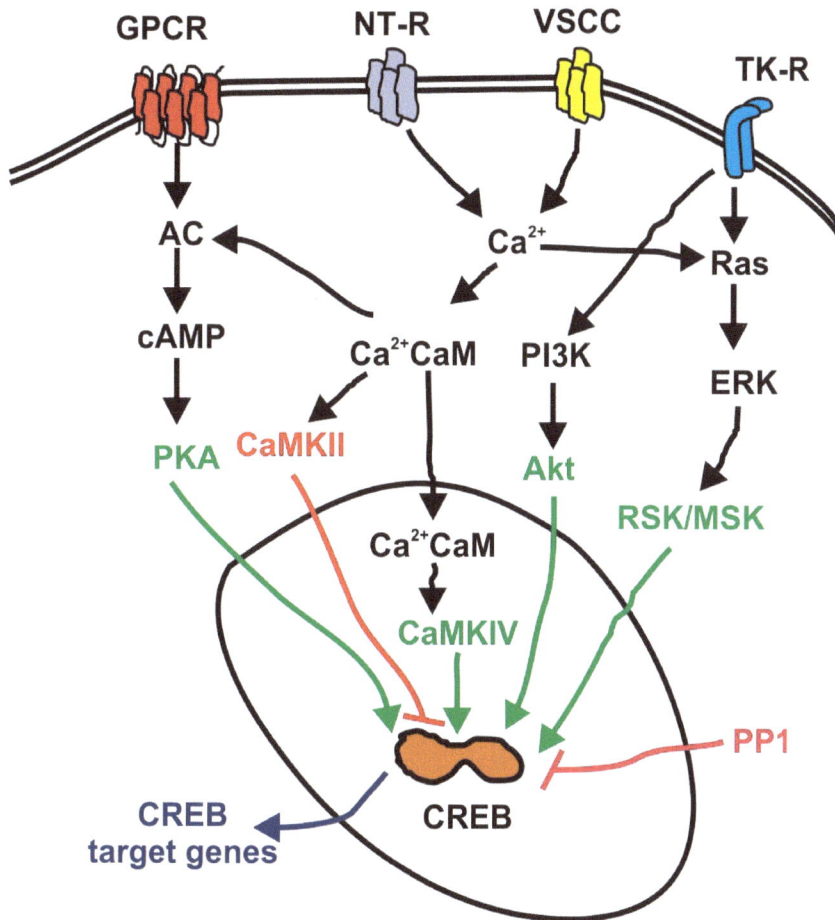

Fig. (3). Overview of signalling pathways that converge at CREB. Excitatory neurotransmitters, ligands for GPCRs and neuronal growth factors are among the stimuli that activate signalling pathways that converge upon CREB. As described in the text, multiple stimulus-dependent protein kinases have been implicated as CREB kinases in neurons, and a high degree of cross-talk exists between these signalling pathways. Stimulus-dependent CREB kinases include PKA, CaMKIV, and members of the RSK and MSK families of protein kinases. Protein phosphatase 1 (PP1) has been implicated as the predominant phospho-CREB phosphatase. Green and red lines represent CREB activation and inhibition respectively.

THE SIGNALLING PATHWAYS AND KINASES THAT LEAD TO CREB PHOSPHORYLATION

i) cAMP Signalling to CREB

Studies of the mechanisms of cAMP-induced somatostatin transcription led to the important discovery of CREB Ser-133 phosphorylation as a modification central to CREB activation. The kinase responsible for this phosphorylation was identified as the cAMP-dependent protein kinase, PKA [55]. PKA activity is regulated by molecules that alter cAMP levels and therefore, by molecules that regulate adenylate cyclase activity, the best characterized of which are the G protein-coupled receptors (GPCRs). In the nervous system, GPCRs function as receptors for functionally important ligands, including many neurotransmitters and neuropeptides. By means of these receptors, neurotransmitters and neuropeptides can couple to cAMP, PKA and ultimately, CREB. However, the ability to activate CREB *via* PKA is not restricted to the ligand activation of GPCRs since some subtypes of adenylate cyclases are regulated by Ca^{2+}. Thus CREB activation *via* PKA can also occur in response to the many of the stimuli capable of increasing intracellular Ca^{2+} [reviewed in 56].

ii) Ca^{2+} Signalling to CREB

Early studies demonstrated that multiple neurotransmitters are capable of activating the expression of IEGs in a Ca^{2+}-dependent manner [for example 30, 57-61] and subsequently, CREB was shown to be able to function as a Ca^{2+}-inducible transcription factor [62, 63, reviewed in 18]. In neurons, increases in intracellular Ca^{2+} occur through voltage- or ligand-gated cation channels and upon membrane depolarization, Ca^{2+} influx occurs *via* voltage-sensitive calcium channels (VSCC) such as L type Ca^{2+} channels. Alternatively, during glutamatergic synaptic transmission glutamate binds to ionotropic receptors like the NMDA subtype, which can function as cation-permeable ion channels when activated. Ca^{2+} interacts with a large number of intracellular molecules, such as the Ca^{2+} binding protein calmodulin (CaM), although it is not clear whether the CaM activation necessary for signalling to the CREB kinase requires cytoplasmic [23] or nuclear Ca^{2+} [64]. Nevertheless, Ca^{2+}-CaM can activate CaMKI, CaMKII and CaMKIV, each of which can phosphorylate CREB *in vitro* at least [62-65]. Of these, CaMKIV has emerged as the most important Ca^{2+}-activated CREB kinase *in vivo* [66-70]. In addition to its ability to activate CaMKs, Ca^{2+} appears to also activate an additional pathway, the Ras/ERK pathway [20, 71-75, reviewed in, 21] that signals to an independent set of CREB kinases.

iii) Growth Factor Signalling to CREB

Neuronal growth factors, like glutamate, appear to be capable of triggering parallel pathways that provoke CREB phosphorylation through the activation of a single receptor [76-78]. Many such growth factors, including neurotrophins, signal *via* receptor tyrosine kinases that activate several known signalling cascades upon ligand binding and dimerization [reviewed in 79, 80]. One such pathway is that mediated by Ras/ERK, a kinase cascade that ultimately results in CREB phosphorylation. Ras/ERK-dependent phosphorylation of CREB may be mediated by several different kinases, including members of the RSK and MSK families. RSK1, RSK2, and RSK3 are MAPK activated kinases each of which has been shown to phosphorylate CREB in cell lines in response to growth factors [81-83]. The structurally related MSK1 and MSK2 protein kinases are also activated by MAPK pathways, and their roles in growth factor-dependent phosphorylation of CREB has recently been assessed genetically [84, 85]. Thus, depending on the cell type and the nature of the growth factor stimulus, one or more of the five related CREB kinases (RSK1-3 and MSK1/2) may catalyze CREB Ser-133 phosphorylation.

In addition to the MAPK pathways, receptor tyrosine kinases activate a second major signalling pathway, the PI3-kinase/Akt pathway [reviewed in 86]. Like the MAPK pathways for which routes to CREB are well documented, there is now evidence indicating that the PI3-kinase/Akt pathway is important for CREB activation, at least under some circumstances. Indeed, Akt can mediate CREB activation and CRE-mediated transcription in cell lines in response to serum or IGF-1 stimulation [87, 88]. Moreover, recent pharmacological studies have implicated the PI3-kinase pathway in the control of CREB phosphorylation in neurons [89, 90]. However, whether Akt can directly phosphorylate CREB remains unknown, and it seems likely that one of the CREB kinases mentioned above catalyzes PI3-kinase and Akt-dependent phosphorylation of CREB. Nevertheless, the evidence available suggests that at least two major receptor tyrosine kinase-activated pathways may contribute to CREB phosphorylation *via* multiple CREB kinases. Interestingly, due to their size and morphological complexity, neurons face a unique challenge in converting extracellular stimuli, such as that mediated by neurotrophins, into nuclear signals that provoke CREB phosphorylation. For peripheral neurons, long range retrograde signalling can be achieved through a mechanism in which activated ligand-receptor complexes are physically transported from distal axons to the cell bodies [91-93].

PHOSPHATASES INVOLVED IN CREB DEPHOSPHORYLATION

To date, two phosphatases, PP1 and PP2A, are thought to be capable of directly dephosphorylating CREB. Both PP1 and PP2A have been implicated in removing the phosphate moiety from Ser-133 [94-96]. PP1 is involved in the dephosphorylation of Ca^{2+}-activated CREB *in vitro* [97] and *in vivo* [98], and it is believed to underlie the transient nature of CREB phosphorylation in response to brief electrical stimuli. However, physiological contexts are now beginning to be identified in which stimulus-dependent activation, and hence the negative regulation of CREB, appears to be important. One exciting example of such negative regulation involves the different subtypes of NMDA receptors, NR2A versus NR2B containing receptors, which differ in their ability to trigger the active dephosphorylation of CREB. When activated, extrasynaptically located NR2B containing NMDA receptors initiate an intracellular signalling pathway that acts as a CREB shut-off pathway [99]. This observation is consistent with the finding that the duration of CREB phosphorylation upon NMDA receptor activation in hippocampal neurons is correlated with developmental maturity, which may reflect developmental changes in the composition of NMDA receptor complexes (NR2A predominates over NR2B subunits) [100]. In both cases, the transient CREB phosphorylation is attributed to dephosphorylation through the activation of a CREB phosphatase. During early neuronal development the phosphatase involves in CREB dephosphorylation seem to be PP1 while in the mature neurons the identity of the phosphatase is not clear. Nevertheless, these findings suggest a provocative mechanism by which different stimuli and different signalling pathways can control the kinetics and duration of CREB phosphorylation. Hence, the differential control of phosphatase activity may represent an important regulatory parameter contributing to the specificity of CREB-dependent gene expression.

REGULATION OF CREB-DEPENDENT GENE EXPRESSION

The Classical View of CREB-Dependent Transcription

The prevailing view is that in unstimulated cells, CREB is found in the nucleus bound to the promoters of CREB-regulated genes (Fig. **4A**) [18]. Indeed, occupancy of CREB target promoter is regulated in a cell-specific manner and the ability of CREB to bind to a particular CRE represents an important component of gene regulation [101]. The core promoter configuration, including the presence of a TATA box motif, and the vicinity to and methylation state of consensus CREs near the promoter, are key factors in defining the proportion of CREB target genes that will respond to cAMP in a given cell type [102, 103]. Thus, before CREB is phosphorylated in response to a given stimulus in a particular cell type, genetic and epigenetic factors determine whether that cell will respond to the stimulus by upregulating CRE transcription simply by allowing or preventing the CREB-controlled machinery to approach a specific promoter.

After cellular stimulation, signalling pathways are activated to transmit signals from the synapse to the nucleus and to promote CREB phosphorylation at Ser-133. CREB is phosphorylated by multiple signal transduction pathways [39, 104, 105] and after Ser-133 phosphorylation, CREB recruits the transcriptional co-activator CBP, or its paralogue P300 (Fig. **4B**) [52, 107]. CBP/P300 recruitment to Ser-133 phosphoCREB is favoured by signalling-dependent modifications of CBP that are induced by several kinases at Ser-301 [107, 108]. The Ca^{2+}-dependent phosphorylation of CBP is therefore an additional check-point controlling CREB/CBP-dependent transcription, although the mechanism by which CBP phosphorylation affects its transactivating properties is not known.

In addition, structure/function analyses identified the constitutive activation domain (Q2/CAD; Fig. (**2**) as the CREB domain responsible for basal transcriptional activity [50, 109]. Consistent with its ability to promote basal levels of gene expression, the CAD was subsequently found to associate either directly or indirectly with several components of the basal transcriptional activation complex, including TFIIB, TAF110, and TBP [51, 110, 111]. Thus, CAD functions independently of KID, although the converse may not be true as CAD appears also to be required for KID to initiate stimulus-dependent gene expression [50].

Taking the functions of KID and CAD together, a simple model of stimulus-induced CREB-mediated transcription can be considered. Hence, in the absence of stimulation, CREB constitutively assembles the basal transcriptional machinery *via* its CAD domain, resulting in a low level of CRE-driven transcription. Upon stimulation and phosphorylation of CREB at Ser-133, CBP is recruited to CREB and to the CRE sequence in the promoter. The recruitment of this co-activator fulfils at least two important functions. First, CBP can bind basal transcriptional components and therefore, it is believed to stabilize the pre-initiation complex that forms at the promoter [106, 112, 113]. Secondly, the endogenous histone acetyl-transferase (HAT) activity of CBP [114] acetylates histones, facilitating the chromatin unravelling and thereby increasing the accessibility of the local chromatin to transcriptional complexes. Through these two activities CBP is thought to increase the transcriptional activity of the CREB complex and CRE promoter activity in a stimulus-dependent manner.

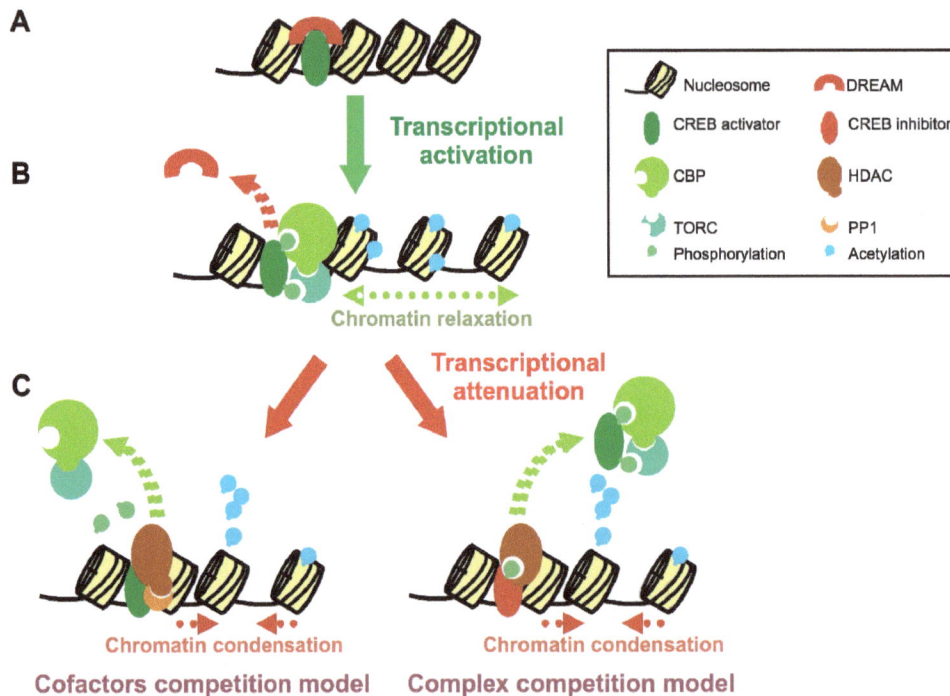

Fig. (4). Tentative model for CREB-dependent transcription activation and attenuation. A, In the basal state CREB is bound to its DNA site. At this moment, CREB interacts with the DREAM protein. B, Extracellular signals are internalized and when they reach the nucleus of the neuron, they provoke transcriptional activation *via* CREB phosphorylation and the release of DREAM. Simultaneously, CBP and TORC are recruited to phospho-CREB and provoke chromatin relaxation by histone acetylation. C, When the extracellular signal disappears, CREB repressors and a co-repressor are recruited to CREB target promoters and they induce chromatin condensation by histone deacetylation.

Additional CREB Phosphorylation Sites Influence CREB-Dependent Transcription

One critical flaw in the model described above is the assumption that Ser-133 phosphorylation is synonymous with CREB-dependent transcription. Several groups have reported that some extracellular stimuli that are at least transiently capable of phosphorylating CREB on Ser-133, fail to induce CREB-dependent gene expression [76, 97, 115-117]. Indeed, the regulation of CREB-dependent gene expression may involve modifications of CREB at sites other than Ser-133 and there are several potential phospho-acceptor sites within CREB [46]. While an early study reported that phosphorylation of CREB at Ser-142 can prevent CREB-mediated transcription [118], the significance or relevance of this event, and of other CREB phospho-acceptor sites, has only recently been revisited. Accordingly, it was shown that the Ser-142 residue is phosphorylated *in vivo* in the mouse supra-chiasmatic nucleus in a circadian-dependent pattern mirroring that of Ser-133 phosphorylation [119]. How Ser-142 might be involved in stimulus-dependent gene expression is not yet clear, although CREB is phosphorylated at Ser-142 and Ser-143 in cultured cortical neurons in a Ca^{2+} specific manner [120]. Indeed, phosphorylation of all three Serine residues (133, 142 and 143) is needed for maximal Ca^{2+}-induced CREB-dependent gene transcription and in agreement with earlier findings [118], phosphorylation at these sites prevents CREB-CBP interactions. The observation that Ser-142/143 phosphorylation may disrupt CREB-CBP binding [120], but that it is required for CREB-CBP dependent gene expression in response to certain stimuli [119, 120], implies that some CREB-dependent genes may be expressed at least partially independent of CBP, or at least independent of a direct CREB-CBP interaction.

TORC (Transducers of Regulated CREB Activity) in CRE-Dependent Transcription

CREB phosphorylation alone is not a reliable predictor of target gene activation and additional regulatory partners for CREB are required to recruit the transcriptional apparatus to the promoter. Calcium signals destabilize the

CREB-CBP complex by promoting Ser-142 phosphorylation [120] and the DNA binding/dimerization domain (bZIP) of CREB mediates a transcriptional response to both calcium influx and cAMP [63]. In this case, bZIP appears to significantly induce CREB activity through its association with a calcium-regulated co-activator. Indeed, a conserved family of co-activators, called TORCs (transducers of regulated CREB activity) was found to interact with CREB and induce CREB activity, enhancing CRE-dependent transcription *via* a phosphorylation-independent interaction with the bZIP domain of CREB. TORC recruitment does not appear to modulate CREB DNA binding activity but rather, it enhances the interaction between the glutamine-rich transactivation domain of CREB and the TAFII130 protein, a component of the TFIID complex [102, 121]. TORC family members share a coiled-coil domain that associates as a tetramer with the bZIP domain of CREB. In basal unstimulated conditions, TORC proteins are sequestered in the cytoplasm through a phosphorylation-dependent interaction with 14-3-3 proteins. Following stimulation, calcium influx increases calcineurin activity, triggering TORC dephosphorylation and its release from 14-3-33 proteins, thereby facilitating its translocation to the nucleus and its interaction with CREB independently of the phosphorylation status (Fig. **4B**) [122-124]. Moreover, recent studies indicate that the interaction of CREB with TORC contributes to CBP/P300 recruitment and enhances its occupancy at relevant CREB target genes [125, 126]. This observation suggests that cooperative interactions between two co-activators lead to the expression of specific genes *via* a ubiquitous transcription factor.

Transcriptional Repressors Modulate CREB Activity

In keeping with the phosphorylation-independent regulation of CREB activity, a Ca^{2+}-dependent interaction between CREB and the transcriptional repressor DREAM (downstream responsive element antagonist modulator), a multifunctional protein of the calcium sensor subfamily [127], regulates the accessibility of CBP to the KID domain in CREB, thereby compromising CRE transcription (Fig. **4A** and **B**) [128]. This interaction requires two leucine rich charged domains (LCDs) present in DREAM and the CREB subfamily proteins [129]. The presence of functional LCDs within DREAM may allow this protein to interact with CREB in a Ca^{2+}-dependent manner, preventing CBP recruitment to phosphoCREB and reducing CRE-dependent transcription without DREAM directly binding to the CRE site [128].

Another important CREB repressor is the ICER protein (inducible cAMP early repressor) [130-132]. There is a family of four ICER isoforms that are produced from an internal p2 promoter located in an intron of the CREM gene [132]. The ICER proteins contain DNA binding/leucine zipper domains that make them endogenous inhibitors of transcription driven by CREB, CREM and ATF1. ICER expression is induced by a variety of stimuli in the brain and in neuronal cultures. Once its expression is induced, ICER homodimers and CREB/ICER heterodimers repress the CREB-dependent transcription by competing with CREB homodimers for the CRE site, or by abolishing CBP recruitment to CREB [133].

Role of Histone Deacetylase (HDAC) in Attenuating CREB-Dependent Transcription

As described above, the Ser-133 residue in CREB plays a key role in CREB transcriptional activation. However, CREB phosphorylation is transiently modulated by recruitment of PP1 to CREB (Fig. **4C**) [94, 95]. It was recently discovered that HDAC inhibitors block CREB dephosphorylation during the attenuation phase of CREB transcription [134]. The fact that HDACs interact with PP1 indicate that both proteins are co-ordinated to prolong CREB phosphorylation and the CBP-CREB association [135, 136]. Alternatively, a competitive association of CREB-CBP and the CREB repressors-HDAC at the c/ebp promoter has been demonstrated in *Aplysia*, during excitatory and inhibitory signal integration (Fig. **4C**) [137]. Together these data indicate that CREB transcriptional activation is finely regulated by co-activators and co-repressors that provoke chromatin remodelling.

CREB TARGET GENES

The ability of CREB to influence the activity of neurons and brain circuits, or complex behaviour, depends on its ability to alter the expression of its target genes. There are now hundreds of genes whose regulatory regions appear to contain CRE or CRE-like sequences (Table **1**), however, the extent to which expression of these genes truly depends upon CREB, CREM or ATF-1 is largely unknown in the majority of cases. Thus, while CREB family members are involved in many processes, identifying the direct target genes is still an important question to resolve. Efforts have been made recently to identify the CREB 'transcriptome' (all those genes regulated by CREB) using chromatin immunoprecipitation and related techniques [102, 103, 138]. In addition, DNA microarrays have been used to identify CREB targets within particular brain regions [139], although much more work is needed to validate the targets identified and to understand their roles under physiological and pathological conditions. Some of the genes identified encode other transcription factors (e.g. c-fos) [61], transcriptional repressors (e.g. cAMP response element modulator, CREM) [45], intracellular messengers (e.g. adenylate cyclase (AC)-VIII) [140], neurotransmitter synthetic enzymes (e.g. tyrosine hydroxylase, TH) [141], peptide transmitters (e.g. corticotropin releasing factor, CRF, and somatostatin) [39, 142], growth factors (e.g. brain-derived neurotrophic factor, BDNF) [77], opioid peptides (dynorphin and enkephalin) [143-145] and neurotransmitter receptor subunits (e.g. GluR1) [146-148]. However, these are likely to represent a small subset of the genes whose regulation by CREB in the brain influences behaviour under normal and pathophysiological conditions.

Table 1.   CREB Gene Targets. The Table Shows a Summary of Some Well Characterized CREB Target Genes Organized into Functional Categories [138, 233].

CREB Target Genes		
Neurotransmission	**Growth factors**	**Structural**
Acetilcholinesterase	BDNF	E-Cadherin
GABA receptor	IGF-I	Fibronectin
Adrenergic receptors	Leptin	ICAM-I
CGRP	TNF-β2	Neurofilament 68kDa
Dopamine β hidroxylase	TrkB	
Enkephalin		**Cellular metabolism**
Cholecystokinin	**Transcription**	Arylakylamine N-acetultransferase
Galanin	c-Fos	Cyclooxigenase 2
Galanin receptor	Egr-1	Cytochrome c
Neurotensin	Icer	Glutamine synthetase
Norepinephrin transporter	JunD	Hemo oxigenase
Preprotachykinin A	Nurr-1	Neuron specific enolase
Prodynorphin	Pit-1	Ornithine decarboxylase
Somatostatin	Stat-3	Phosphoenolpyruvato decarboxykinase
Somatostatin receptor		Pyruate carboxyase
Susbtance P receptor	**Signal transduction**	Superoxido dismutase 2
Synapsin	Cyclin A	Ubiquitin conjugating enzymes
Vasopresin	Cyclin D	Uncoupling protein 1
Vesicular monoamine transporter VGF	i-Nos	Uncoupling protein 2
	Neurofibromastosis	Uncoupling protein 3
	Postgandin cyclise	
	Postagladin synthase	
	SGK	

THE ROLE OF CREB IN LEARNING, MEMORY AND SYNAPTIC PLASTICITY

CREB is Required for Long-Term Plasticity in Invertebrates

Using pharmacological inhibitors, early behavioural studies of learning and long-term memory revealed a requirement for new protein synthesis and new gene transcription in these processes. Subsequent studies demonstrated that this requirement also applies to long-term changes in synaptic plasticity, the cellular correlate of memory [reviewed in 149, 150]. A series of breakthrough studies were then carried out using the mollusc *Aplysia*, which exhibits memory-like behaviour (the siphon withdrawal reflex) known as sensitization and that can be conveniently recapitulated in culture in a paradigm known as long-term facilitation (LTF) [reviewed in 149, 151]. The duration of this response reflects memory retention is a function of the number and intensity of the stimulus received [152]. Thus, while short-term sensitization induced by a weak stimulus lasts for a few minutes and depends on post-translational modifications, multiple stimuli induce sensitization that lasts for weeks and requires new protein and RNA synthesis. Short- and long-term sensitization are accompanied by short- and long-term synaptic facilitation of neurotransmitter release, respectively. This system provided the first mechanistic insight into the role of activity-dependent gene expression in learning and memory, demonstrating a requirement for the cAMP-CREB pathway [155, reviewed in 149]. Moreover, it paved the way for a series of elegant studies demonstrating that the CREB transcriptional pathway is both necessary and sufficient for LTF [156-159, reviewed in 149]. Furthermore, not only PKA but also mitogen-activated protein kinase (MAPK or ERK) is essential in activating the CREB pathway. In fact, in the *Aplysia* model it has been proposed that after five pulses of 5-HT, activated PKA and MAPKs translocate to the nucleus where they presumably phosphorylate an activate CREB1, an activator form of CREB [153, 154]. In parallel, CREB2, a repressor form of CREB, is removed and hence, gene expression is activated [158].

The fruitfly, *Drosophila melanogaster*, has also been exploited for learning and memory studies. The development of the olfactory memory paradigm as a simple behavioural assay for learning and memory in *Drosophila* enabled geneticists to screen for mutant genes that disrupt these processes [reviewed in 160]. Remarkably, these genes turned out to include key regulators of intracellular cAMP levels. In particular, two genes, *Rutabaga* and *Dunce*, produce profound learning and memory defects when disrupted and they were found to encode for a Ca^{2+}/CaM-sensitive adenylate cyclase and a phosphodiesterase, respectively [reviewed in 160]. CREB itself was first implicated in learning and memory in flies when expressing a CREB activator or repressor was found to enhance or block the formation of long-term memory in an olfactory task [161, 162].

CREB Modulates Learning and Memory in Vertebrates

These data from *Aplysia* and *Drosophila* imply that CREB plays a role in learning and memory in invertebrates and as a result, much subsequent effort has been devoted to determine whether the same holds true for vertebrates. The first studies in rodents supported a role for CREB in learning, memory and synaptic plasticity, since manipulating multiple steps in the cAMP-PKA pathway altered these processes. First, mice with a targeted mutation in the type I adenylate cyclase were found to have defects in both spatial memory and L-LTP in the CA1 region of the hippocampus [163]. Furthermore, cAMP analogues [164] and antagonists [165] were found to have opposing effects on hippocampal LTP in slice cultures: while the former triggers LTP in the absence of a tetanizing stimulus, the latter blocks L-LTP. Finally, in mice in which the activity of the cAMP effector PKA was blocked by means of a transgene expressing a PKA inhibitor, both spatial learning and L-LTP were impaired [166]. Given that multiple components of the cAMP-PKA pathway are involved in learning and memory, it is reasonable to suspect that CREB-dependent gene expression would also be activated in this process. Indeed, robust CREB phosphorylation and CRE-reporter gene expression can be detected in cortical neurons during developmental plasticity [167], and in hippocampal neurons in response to both LTP-inducing stimuli and memory training tasks [20, 168-174]. CREB-dependent gene expression appears to be a requisite rather than a consequence of learning and memory since intra-hippocampal infusion of CREB antisense oligonucleotides produces deficits in spatial learning in rats [175]. In a related study, CREB signalling was reported to be necessary for plasticity in an *in vitro* model of cerebellar LTD [176]. These findings support a model in which CREB-dependent gene expression contributes critically to long-term memory and plasticity in vertebrates, and they provided the impetus to attempt to directly manipulate CREB levels *in vivo*. However, curiously such genetic experiments have ultimately led to a series of reports demonstrating the great complexity and contradictions in some circumstances.

The first genetic tool to study the physiological role of CREB was a hypomorphic mutation in the *Creb* gene [177]. These mice fail to express the predominant forms of CREB, α and Δ, but trace amounts of other CREB proteins are produced that can fulfil some CREB-mediated functions *in vivo* [178]. The first behavioural study using these mice reported deficits in both spatial memory and in hippocampal LTP, suggesting that as in invertebrates, CREB may be required for long-term memory in mice [179]. However, more recent studies imply that CREB is either dispensable for certain forms of hippocampal plasticity, or that compensation by other family members may overcome the loss of CREB *per se in vivo* [180-182]. Other studies using a transgenic mice with regulated expression of the CREB-VP16 fusion protein, a constitutively active CREB protein, showed improved learning and memory and long-LTP in mice [183, 184], although neuronal death was caused by overexcitability [185, 186]. Alternatively, CREB inhibition in mice supported the notion that CREB is involved in certain forms of memory, even though it is not absolutely requirement for plasticity. Indeed, inducible expression of a CREB inhibitor blocks consolidation of long-term fear-conditioned fear memories [187]. Moreover, inducible expression of a dominant inhibitor of all CREB family members in the dorsal hippocampus produces spatial memory deficits [182]. Interestingly, the latter study also revealed a differential requirement for CREB family members in the expression of different types of LTP. While the CREB family inhibitor had no effect on L-LTP induced either by theta bursts or tetanic trains, it did attenuate L-LTP induced by cAMP, and by the pairing of electrical stimulation with neurotransmitter application [182]. Furthermore, when an artificial peptide with a strong and broad inhibitory effect on the CREB family, A-CREB, is expressed in the hippocampus in an inducible manner, long-LTP is impaired while intrinsic excitability and the susceptibility to induced seizures are reduced, and both basal and activity-driven gene expression are altered [188].

Table 2.   **CREB Co-Factors Modulate Neuronal Plasticity. The Table Shows a Summary of the Cognitive Behavioural and Electrophysiological Alterations Associated with Genetic Modification of CREB Co-Factors in Mice.**

CREB Cofactor	Genetic Manipulation	Physiological Effects
DREAM [189, 190]	Gene deletion	Learning facilitation Longer memory CREB transcription threshold decrease
ICER [191]	Gene deletion	Longer memory
	Gene overexpression	Memory disfunction
PP1 [196]	Overexpression of inhibitor peptide	Longer memory
CBP [revised in 195]	Heterozygote gene deletion Heterozigote CBP truncation	Impairment in long-term memory in object recognition and fear conditioning Long-LTP impairment
	Conditional CBP mutated in HAT domain	Impairment in long-term memory in object recognition and water maze
	Postnatal expression of CBP mutated	Impairment in long-term memory in fear conditioning and water maze Impairment in Long-LTP induced by DA
HDAC2 [197]	Gene deletion	Increase in synapsis number Memory facilitation
	Gene overexpression	Decrease in synapsis number Memory impairment

CREB activity is modulated by co-factors and by modifying their activity, CREB function can also be examined (Table 2). Recent studies have demonstrated that mice lacking the CREB repressors DREAM or ICER had improved cognitive capacities. For example, mice lacking DREAM presented learning facilitation and long-lasting memories

[189, 190], while manipulation of ICER in mice provoked memory enhancement in its absence or memory impairment when it is overexpressed [191]. By contrast, expression of a dominant negative version of TORC blocked long-term synaptic plasticity [192-194]. When CBP is modified, memory and LTP are impaired in mice, although the severity depends on the mutation expressed and the time course of expression [195]. Finally, deletion of phosphatase 1 (PP1) or HDAC2 is involved in CREB transcription attenuation, provoking memory enhancement [98,196,197]. All these studies confirm the central role of CREB activation in synaptic plasticity events related with learning and memory processes. Moreover, individual cognitive capacities may be determined by the interaction between CREB and CREB co-factors, which are responsible for establishing the CREB activation-deactivation equilibrium.

From all these studies, the following question arises: how long and in which brain regions is CREB critically involved in memory formation? Bilateral infusion of CREB antisense ODN into the rat dorsal hippocampus showed that in this region, CREB-mediated transcription is necessary for learning and remembering spatial tasks [175]. Moreover, this requirement is temporally limited to the post-training phase, suggesting that the CREB-dependent gene expression required for memory consolidation is relatively brief. In parallel, biochemical studies of the active and inactive state of CREB after training for different types of learning tasks helped determine the temporal dynamics of CREB activation underlying memory formation. Accordingly, the learning-specific CREB phosphorylation increases in brain areas related with learning consolidation in an inhibitory avoidance task and in a contextual fear conditioning paradigm following a biphasic profile [172, 198]. In both cases, the first peak of activation commenced immediately after training, returning to baseline after 30 min, with a second peak arising between 3 and 6 h post-training that returned to baseline by 9 h after training. This pattern of CREB activation seems to be critical for establishing long-term memory, as demonstrated by the fact that administering phosphatase inhibitor impairs learning consolidation of a passive avoidance task in chicks [199]. Interestingly, LTP is accompanied by biphasic activation of CREB *in vivo* [171, 200] and although the LTP protocols appear to elicit a first peak of CREB phosphorylation after 30 min, the second peak arises over a more variable timescale. Studies in rat organotypic hippocampal slices also revealed that the induction of LTP in CA1 is accompanied by a local increase in CREB phosphorylation [201] that is sustained for at least 4 h, again supporting the hypothesis that CREB is important in the late phase of LTP. Likewise, changes in CREB function, which probably result from coincidental integration of inputs, influence the probability that individual neurons will be recruited to encode memory. Accordingly, only about one-quarter of neurons that receive sensory inputs show CREB phosphorylation in the lateral amygdala after auditory fear conditioning, indicating that CREB activation is critical to determine which neurons are recruited into the fear memory trace [202]. In other words, memory formation seems to recruit eligible neurons in function of their relative CREB activity at the time of learning.

If CREB and its family members are both necessary and sufficient to induce synaptic modifications, how can activity at synaptic sites activate CREB in the nucleus such that its outputs could be restricted to the activated synapse? The most provocative model is that of "synaptic tagging" or the "synaptic capture" model, whereby plasticity-inducing stimuli "activate" the postsynaptic neuron and the nuclear output of this activation is captured by tags at active synaptic sites, including those that receive weaker inputs. Support for this model has been obtained from studies of the synapse specificity of LTF in invertebrate sensory neurons [203, 204] and of LTP in vertebrate hippocampal neurons [205]. Such a model provides a mechanism to explain how subthreshold stimulation of a given synapse can result in LTP at that synapse if LTP had previously been induced at a separate synapse [205]. Indeed, it was proposed that a central activating event that follows the LTP-inducing stimulus is the activation of CREB [183]. Thus, while CREB alone may not be absolutely required for the expression of LTP, the products of its target genes appear to provide sufficient signals for synaptic strengthening. The identities of these CREB target genes and the mechanisms by which their protein products are captured remain to be addressed.

THE ROLE OF CREB IN NEUROPSYCHIATRIC DISEASES

CREB affects learning and memory because it is a crucial mediator of experience-based neuroadaptation. Experience can take many forms, such as exposure to a maze, to a physical or emotional stress, or to an addictive substance. Each of these stimuli can affect the intracellular pathways that converge upon CREB. In turn, alterations in CREB activity fundamentally affect the way that cells function in isolation, within groups, or within larger circuits. Because CREB is found in all neural circuits and it influences the expression of diverse genes, it is not surprising that the consequences of boosting CREB-mediated gene regulation are sometimes beneficial, while at other times they seem to be detrimental, as may be the case in depression and anxiety disorders [revised in 206, 207].

CREB in Depression

Major depressive disorder is a severe clinical problem worldwide, with a lifetime risk of 10%-30% for women and 7%-15% for men. The World Health Organization ranks major depression at the top of the list in terms of disease burden, and this burden is expected to rise in the next decade as the prevalence of the disorder grows. It is possible that the transcription factor CREB could serve as a point of convergence for multiple classes of antidepressant drugs. Induction of CREB causes very different functional effects, either antidepressant or depressive, depending on the brain region involved [208-210]. Selective overexpression of CREB in the dentate gyrus of the hippocampus has an antidepressant effect on multiple behavioural parameters [211]. By contrast, overexpression of CREB in the amygdala or nucleus accumbens (NAc) has a depressive effect, whereas dominant-negative CREB (mCREB) expression provokes an antidepressant-like response in several behavioural parameters [212-214]. These results highlight the importance of CREB in this type of plasticity, and how the changes evoked by such stimuli differ throughout the brain.

Although the mechanisms responsible for these effects remain under intense study, evidence is accumulating to suggest that CREB may regulate the expression of neural growth factors [215, 216]. In fact, many antidepressant treatments in humans also increase the expression of BDNF, a CREB-regulated target gene, in the hippocampus [217]. Increased growth factor activity within the hippocampus could stimulate regenerative processes such as dendrite sprouting and neurogenesis, each of which are observed after sustained antidepressant treatment in humans [218]. These processes could help restore a normal mood by repairing stress-related damage, helping to explain why the therapeutic effects of antidepressant treatments often take weeks or months to become apparent. Although the hippocampus is certainly not the only brain region involved in depression, it is a prominent limbic structure where abnormalities have been reported in human patients [218, 219]. On the other hand, regulating depression-like behaviour by altering CREB activity within the NAc appears to be partly mediated by dynorphin, an endogenous ligand of k opioid receptors [145]. Administration of selective k-receptor agonists augments depressive behaviour [220, 221], while k receptor selective antagonists attenuate the depressive consequences of elevated CREB expression within the NAc [145, 212]. Although the mechanisms by which k receptor selective antagonists produce antidepressive effects in rats are unclear [212, 220-222], one possibility is that these agents block the k opioid receptors that normally inhibit neurotransmitter release from mesolimbic dopaminergic neurons [223-225]. Considered together, these data raise the possibility that CREB-mediated transcription of dynorphin within the NAc decreases dopamine function, which triggers certain features of depression. Until recently, the NAc has not been considered a likely site for the pathophysiology of depression, although the involvement of this brain reward region would explain the symptoms of anhedonia, reduced energy and reduced motivation that are prominent in many depressed patients [212, 219]. Within the amygdala, the consequences of altering CREB function in models of depression appear to be state-dependent. Overexpression of CREB in the amygdala before training in the learned helplessness paradigm drives depression, whereas expression after training results in antidepressant-like effects [214]. In either case, these findings provide further evidence that the actions of CREB are regionally and temporally specific.

Such observations highlight the fact that CREB is generally involved in regulating plasticity, a process that is not inherently good or bad but that could be adaptive, maladaptive, or both simultaneously. In the case of depression, the seemingly paradoxical requirements for antidepressant efficacy – elevations in CREB activity in one region and reductions in another – could detract from the therapeutic actions of treatments that have a global influence on CREB function in the brain.

CREB in Anxiety

Anxiety is a common human emotion that diminishes quality of life and that increases social burden worldwide. Increasing evidence indicates that CREB can influence the extended amygdala – a macrostructure that includes components of the amygdala, the bed nucleus of the stria terminalis (BNST) and the NAc [226] – and that it can regulate anxiety like behaviours in rats. There is evidence that disrupting CREB function in the NAc produces anxiety-like effects [227, 228], whereas induction of CREB function within the amygdala produces similar behavioural effects [229]. Studies using drugs or CREB knock-out mice also suggest a role for CREB in anxiety disorders, although the conclusions are somewhat different [229, 230], which might be due to differences in the overall patterns of CREB expression throughout the brain. Additional research on the relationship between CREB, CREB-regulated target genes (e.g. neuropeptide Y [229]) and anxiety behaviour might ultimately reveal potential benefits of altering CREB function in the treatment of anxiety-related disorders.

The notion that elevated CREB function in the NAc causes certain depressive-like symptoms, whereas reduced CREB function in this region causes anxiety-like behaviour, might seem paradoxical, although it could be understood that CREB could have a role in this reward circuit under normal conditions. In this sense, CREB is a key regulator of the reactivity of brain reward circuits and thus, it regulates sensitivity to emotional stimuli. Together, these data highlight the notion that extreme variation in CREB function in the NAc might be detrimental, arguing for further caution in designing CREB-based therapies.

CONCLUSIONS

Increases in CREB function can clearly enhance memory under certain circumstances in diverse species. Such data have generated great enthusiasm for the development of therapeutic agents that could improve memory deficits associated with cognitive disorders, such as Alzheimer's disease. Safe and effective cognitive enhancers that would be effective in healthy people would also be of great interest in cultures that are preoccupied with productivity and self-improvement. However, increases in CREB function can also disrupt cognitive performance under some circumstances. The brain regions in which CREB acts to enhance or disrupt memory might be different and it remains to be determined which effect would predominate if it were possible to enhance CREB function throughout the brain. Importantly, CREB also influences other experience-dependent processes that can affect complex motivational behaviour. Some examples of these processes seem to be beneficial, whereas others are detrimental. These effects can often be linked, at least in part, to particular brain regions, highlighting the fact that general alterations in CREB function would not be expected to produce uniform effects throughout the brain. However, defining the links between such effects and brain regions might be crucial if the promise of CREB-related therapeutic drugs is to become a reality in the future. The molecular elements responsible for CREB's specific role in each brain region and neural circuit could potentially be utilized and exploited therapeutically to alter the activity of this transcription factor. Future work could reveal whether certain signal transduction pathways are more important in some brain regions than in others. Another approach might also be to focus on the regulation of downstream targets of CREB [231]. In this sense, a recent study, combining microarray and chromatin immunoprecipitation technology to analyze CREB-DNA interactions in brain, has observed substantial differences in CREB occupancy and phosphorylation between frontal cortex, hippocampus, and striatum that could explain the diversified consequences of CREB activation in these regions [232]. For example, specific CREB targets to hippocampus are enriched in genes involved in transcriptional activation, protein biosynthesis, and synaptic transmission relative to the other regions, while CREB targets unique to striatum are enriched in genes involved in immunity, inflammation, apoptosis, proteolysis, serotonin degradation, and steroid metabolism. These differences likely contribute to the divergent behavioural consequences of enhanced CREB activity in these regions, namely the antidepressant or prodepressive effects of CREB hyperactivity in hippocampus or striatum, respectively [211-213]. Thus, therapies those directly act on specific CREB target genes, and which themselves are expressed more narrowly within affected brain regions, might enable better control over the proteins that cause disease states or their symptoms, while sparing other crucial functions. Regardless, research on CREB is improving our understanding of the molecular mechanisms controlling both normal complex behaviour and the abnormal behaviour associated with prominent psychiatric and neurological conditions, and such efforts might one day aid the discovery of therapies for these disorders.

ACKNOWLEDGEMENTS

We thank Dr Sefton for critical reading and editorial assistance with the manuscript.. This work was supported by grants from the AECI, the Junta de Andalucía (BIO-122), the DGICYT (BFU2008-01552), and the Fundación Ramón Areces.

REFERENCES

[1]   Catterall WA, Few AP. Calcium channel regulation and presynaptic plasticity. Neuron 2008; 59: 882-901.

[2]   Higley MJ, Sabatini BL. Calcium signaling in dendrites and spines: practical and functional considerations. Neuron 2008; 59: 902-13.

[3]   Wayman GA, Lee Y-S, Tokumitsu H, Silva A, Soderling TR. Calmodulin-kinases: modulation of neuronal development and plasticity. Neuron 2008; 59: 914-31.

[4]   Greer PL, Greenberg ME. From synapse to nucleus: calcium-dependent gene transcription in the control of synapse development and function. Neuron 2008; 59: 846-60.

[5]   Hong EJ, West AE, Greenberg ME. Transcriptional control of cognitive development. Curr Opin Neurobiol 2005; 15: 21-8.

[6]   Agranoff BW, Davis RE, Brink JJ. Memory fixation in the gold fish. Proc Nat Acad Sci USA 1965; 54: 788-93.

[7]   Davis HP, Squire LR. Protein synthesis and memory: A review. Psychol Bull 1984; 96: 518-19.

[8]   Goelet P, Castellucci VF, Schacher S, Kandel ER. The long and the short of long-term memory—A molecular framework. Nature 1986; 322: 419-22.

[9]   Clayton DF. The genomic action potential. Neurobiol Learn Mem 2000; 74: 185-216.

[10]   Guzowski JF. Insights into immediate-early gene function in hippocampal memory consolidation using antisense oligonucleotide and fluorescent imaging approaches. Hippocampus. 2002; 12: 86-104.

[11]   Tischmeyer W, Grimm R. Activation of immediate early genes and memory formation. Cell and Mol Life Sci 1999; 55: 564-74.

[12]   Abraham WC, Mason SE, Demmer J, *et al.* Correlations between immediate early gene induction and the persistence of long-term potentiation. Neurosci 1993; 56: 717-27.

[13]   Worley PF, Bhat RV, Baraban JM, Erickson CA, McNaughton BL, Barnes CA. Thresholds for synaptic activation of transcription factors in hippocampus: Correlation with long-term enhancement. J Neurosci 1993; 13: 4776-86.

[14]   Adams JP, Dudek SM. Late-phase long-term potentiation: Getting to the nucleus. Nat Rev Neurosci 2005; 6: 737-43.

[15]   Berridge MJ. Neuronal calcium signaling. Neuron 1998; 21: 13-26.

[16]   Cohen S, Greenberg ME. Communication between the synapse and the nucleus in neuronal development, plasticity, and disease. Annu Rev Cell Dev Biol 2008; 24: 183-209.

[17]   Delghandi MP, Johannessen M, Moens U. The cAMP signalling pathway activates CREB through PKA, p38 and MSK1 in NIH 3T3 cells. Cell Signaling 2005; 17: 1343-51.

[18]   Shaywitz AJ, Greenberg ME. CREB: a stimulus induced transcription factor activated by a diverse array of extracellular signals. Annu Rev Biochem 1999; 68: 821-61.

[19]   Davis S, Laroche S. Mitogen-activated protein kinase/extracellular regulated kinase signalling and memory stabilization: A review. Genes, Brain and Behavior 2006: Suppl 2: 61-72.

[20]   Davis S, Vanhoutte P, Pages C, Caboche J, Laroche S. The MAPK/ERK cascade targets both Elk-1 and cAMP response element-binding protein to control long-term potentiation dependent gene expression in the dentate gyrus in vivo. J. Neurosci 2000; 20: 4563-72.

[21]   Sweatt JD. The neuronal MAP kinase cascade: a biochemical signal integration system subserving synaptic plasticity and memory. J Neurochem 2001; 76: 1-10.

[22]   Wu GY, Deisseroth K, Tsien RW. Spaced stimuli stabilize MAPK pathway activation and its effects on dendritic morphology. Nat Neurosci 2001; 4: 151-58.

[23]   Deisseroth K, Heist EK, Tsien RW. Translocation of calmodulin to the nucleus supports CREB phosphorylation in hippocampal neurons. Nature 1998; 392: 198-202.

[24]   Ginty DD. Calcium regulation of gene expression: Isn't that spatial? Neuron 1997; 18: 183-86.

[25]   Soderling TR, Chang B, Brickey D. Cellular signalling through multifunctional Ca^{2+}/calmodulin dependent protein kinase II. J Biol Chem 2001; 276: 3719-22.

[26]   Finkbeiner S, Greenberg ME. Ca2+ channel-regulated neuronal gene expression. J Neurobiol 1998; 37: 171-89.

[27]   Graef IA, Mermelstein PG, Stankunas K, *et al.* L-type calcium channels and GSK-3 regulate the activity of NF-ATc4 in hippocampal neurons Nature 1999; 401: 703-8.

[28]   Meffert MK, Chang JM, Wiltgen BJ, Fanselow MS, Baltimore D. NF-kappa B functions in synaptic signaling and behaviour. Nat Neurosci 2003; 6: 1072-78.

[29]   Alberini CM. Transcription factors in long-term memory and synaptic plasticity. Physiol Rev 2009; 89: 121-45.

[30]   Greenberg ME, Ziff EB, Greene LA. Stimulation of neuronal acetylcholine receptors induces rapid gene transcription. Science 1986; 234: 80-3.

[31]   Guzowski JF, McNaughton BL, Barnes CA, Worley PF. Environment-speci.c expression of the immediate-early gene Arc in hippocampal neuronal ensembles. Nat Neurosci 1999; 2: 1120-24.

[32]   Vazdarjanova A, McNaughton BL, Barnes CA, Worley PF, Guzowski JF. Experience-dependent coincident expression of the effector immediate-early genes arc and Homer 1a in hippocampal and neocortical neuronal networks. J Neurosci 2002; 22: 10067-71.

[33]   Lanahan A, Worley P. Immediate-early genes and synaptic function. Neurobiol Learn Mem 1998; 70: 37-43.

[34]   Nedivi E, Hevroni D, Naot D, Israeli D, Citri Y. Numerous candidate plasticity-related genes revealed by differential cDNA cloning. Nature 1993; 363: 718-22.

[35]   Morgan JI, Curran T. Stimulus-transcription coupling in the nervous system: Involvement of the inducible proto-oncogenes fos and jun. Ann Rev Neurosci 1991; 14: 421-51.

[36]   Shepherd JD, Rumbaugh G, Wu J, *et al.* Arc/Arg3.1 mediates homeostatic synaptic scaling of AMPA receptors. Neuron 2006; 52: 475-84.

[37]   Montminy M. Transcriptional regulation by cyclic AMP. Annu Rev Biochem 1997; 66: 807-22,

[38]   De Cesare, D., Fimia, G.M. Sassone-Corsi, P. Signaling routes to CREM and CREB: plasticity in transcriptional activation. Trends Biochem Sci 1999; 24: 281-85.

[39]   Mayr, B, Montminy M. Transcriptional regulation by the phosphorylation-dependent factor CREB. Nat Rev Mol Cell Biol 2001; 2: 599-609.

[40]   Meyer TE, Habener JF. Cyclic adenosine 3-,5'-monophosphate response element binding protein (CREB) and related transcription-activating deoxyribonucleic acid-binding proteins. Endocr Rev 1993; 14: 269-90.

[41]   Montminy MR, Bilezikjian LM. Binding of a nuclear protein to the cyclic-AMP response element of the somatostatin gene. Nature 1987; 328: 175-78.

[42]   Hurst HC, Masson N, Jones NC, Lee KA. The cellular transcription factor CREB corresponds to activating transcription factor 47 (ATF-47) and forms complexes with a group of polypeptides related to ATF-43. Mol Cell Biol 1990; 10: 6192-203.

[43]   Nichols M, Weih F, Schmid W, *et al*. Phosphorylation of CREB affects its binding to high and low affinity sites: implications for cAMP induced gene transcription. EMBO J 1992; 11: 3337-46.

[44]   Bohm S, Bakke M, Nilsson M, Zanger UM, Spyrou G, Lund JJ. Cooperating nonconsensus cAMP-responsive elements are mediators of adrenocorticotropin-induced VL30 transcription in steroidogenic adrenal cells. J Biol Chem 1993; 268: 3952-63.

[45]   Foulkes NS, Borrelli E, Sassone-Corsi P. CREM gene: use of alternative DNA-binding domains generates multiple antagonists of cAMP-induced transcription. Cell 1991; 64: 739-49.

[46]   Gonzalez GA, Yamamoto KK, Fischer WH, *et al*. Cluster of phosphorylation sites on the cyclic AMP-regulated nuclear factor CREB predicted by its sequence. Nature 1989; 337: 749-52.

[47]   Hai TW, Liu F, Coukos WJ, Green MR. Transcription factor ATF cDNA clones: an extensive family of leucine zipper proteins able to selectively form DNA-binding heterodimers. Genes Dev 1989; 3: 2083-90.

[48]   Hoeffler JP, Meyer TE, Yun Y, Jameson JL, Habener JF. Cyclic AMP-responsive DNA-binding protein: structure based on a cloned placental cDNA. Science 1988; 242: 1430-33.

[49]   Ferreri K, Gill G, Montminy M. The cAMP-regulated transcription factor CREB interacts with a component of the TFIID complex. Proc Natl Acad Sci USA 1994; 91: 1210-13.

[50]   Quinn PG. Distinct activation domains within cAMP response element-binding protein (CREB) mediate basal and cAMP-stimulated transcription. J Biol Chem 1993; 268: 16999-17009.

[51]   Xing L, Gopal VK, Quinn PG. cAMP response element-binding protein (CREB) interacts with transcription factors IIB and IID. J Biol Chem 1995; 270: 17488-93.

[52]   Chrivia JC, Kwok RP, Lamb N, Hagiwara M, Montminy MR, Goodman RH. Phosphorylated CREB binds specifically to the nuclear protein CBP. Nature 1993; 365: 855-59.

[53]   Johannessen M, Delghandi MP, Moens U. What turns CREB on? Cell Signal 2004; 16: 1211-27.

[54]   Kalkhoven E. CBP and p300: HATs for different occasions. Biochem. Pharmacol 2004; 68: 1145-55.

[55]   Gonzalez GA Montminy MR. Cyclic AMP stimulates somatostatin gene transcription by phosphorylation of CREB at serine 133. Cell 1989; 59: 675-80.

[56]   Poser S, Storm DR. Role of Ca2_-stimulated adenylyl cyclases in LTP and memory formation. Int J Dev Neurosci 2001; 19: 387-94.

[57]   Bading H, Ginty DD, Greenberg ME. Regulation of gene expression in hippocampal neurons by distinct calcium signaling pathways. Science 1993; 260: 181-86.

[58]   Bartel DP, Sheng M, Lau LF, Greenberg ME. Growth factors and membrane depolarization activate distinct programs of early response gene expression: dissociation of fos and jun induction. Genes Dev 1989; 3: 304-13.

[59]   Morgan JI, Curran T. Role of ion flux in the control of c-fos expression Nature 1986; 322: 552-5.

[60]   Morgan JI, Cohen DR, Hempstead JL, Curran T. Mapping patterns of c-fos expression in the central nervous system after seizure. Science 1987; 237: 192-97.

[61]   Sheng M, McFadden G, Greenberg ME. Membrane depolarization and calcium induce c-fos transcription *via* phosphorylation of transcription factor CREB. Neuron 1990; 4: 571-82.

[62]   Dash PK, Karl KA, Colicos MA, Prywes R Kandel ER. cAMP response element-binding protein is activated by Ca^{2+}/calmodulin- as well as cAMP-dependent protein kinase. Proc Natl Acad Sci USA 1991; 88: 5061-5.

[63]   Sheng M, Thompson MA, Greenberg ME. CREB: a Ca^{2+}-regulated transcription factor phosphorylated by calmodulin-dependent kinases. Science 1991; 252: 1427-30.

[64]   Hardingham GE, Arnold FJ, Bading H. Nuclear calcium signaling controls CREB-mediated gene expression triggered by synaptic activity. Nat Neurosci 2001; 4: 261-7.

[65]   West AE, Chen WG, Dalva MB, *et al*. Calcium regulation of neuronal gene expression. Proc Natl Acad Sci USA 2001, 98: 11024-31.

[66]   Anderson KA, Means AR. Defective signaling in a subpopulation of CD4(+) T cells in the absence of Ca^{2+}/calmodulin-dependent protein kinase IV. Mol Cell Biol 2002; 22: 23-9.

[67]   Ho N, Liauw JA, Blaeser F, *et al*. Impaired synaptic plasticity and cAMP response element-binding protein activation in Ca^{2+}/calmodulin-dependent protein kinase type IV/Gr-deficient mice. J Neurosci 2000; 20: 6459-72.

[68]   Kang H, Sun LD, Atkins CM, Soderling TR, Wilson MA, Tonegawa S. An important role of neural activity-dependent CaMKIV signaling in the consolidation of long-term memory. Cell, 2001; 106: 771-83.

[69]   Ribar TJ, Rodriguiz RM, Khiroug L, Wetsel W, Augustine GJ, Means AR. Cerebellar defects in Ca^{2+}/calmodulin kinase IV-deficient mice. J Neurosci 2000; 20: RC107.

[70]   Wei F, Qiu CS, Liauw J, Robinson DA, Ho N, Chatila T, Zhuo M. Calcium calmodulin-dependent protein kinase IV is required for fear memory. Nat Neurosci 2002; 5: 573-79.

[71]   Bading H, Greenberg ME. Stimulation of protein tyrosine phosphorylation by NMDA receptor activation. Science 1991; 253: 912-14.

[72]   Dolmetsch RE, Pajvani U, Fife K, Spotts JM, Greenberg ME. Signaling to the nucleus by an L-type calcium channel-calmodulin complex through the MAP kinase pathway. Science 2001; 294: 333-9.

[73]   Impey S, Obrietan K, Wong ST, *et al*. Cross talk between ERK and PKA is required for Ca^{2+} stimulation of CREB-dependent transcription and ERK nuclear translocation. Neuron 1998; 21: 869-83.

[74]   Roberson ED, English JD, Adams JP, Selcher JC, Kondratick C, Sweatt JD. The mitogen-activated protein kinase cascade couples PKA and PKC to cAMP response element binding protein phosphorylation in area CA1 of hippocampus. J Neurosci 1999; 19: 4337-48.

[75]   Rosen LB, Ginty DD, Weber MJ, Greenberg ME. Membrane depolarization and calcium influx stimulate MEK and MAP kinase *via* activation of Ras. Neuron 12; 1994: 1207-21.

[76]   Bonni A, Ginty DD, Dudek H, Greenberg ME. Serine 133-phosphorylated CREB induces transcription *via* a cooperative mechanism that may confer specificity to neurotrophin signals. Mol Cell Neurosci 1995; 6: 168-83.

[77]   Finkbeiner S, Tavazoie SF, Maloratsky A, Jacobs KM, Harris KM, Greenberg ME. CREB: a major mediator of neuronal neurotrophin responses. Neuron 1997; 19: 1031-47.

[78]   Ginty DD, Bonni A, Greenberg ME. Nerve growth factor activates a Ras-dependent protein kinase that stimulates c-fos transcription *via* phosphorylation of CREB. Cell 1994; 77: 713-25.

[79]   Patapoutian A, Reichardt LF. Trk receptors: mediators of neurotrophin action. Curr Opin Neurobiol 2001; 11: 272-80.

[80]   Sofroniew MV, Howe CL, Mobley WC. Nerve growth factor signaling, neuroprotection, and neural repair. Annu Rev Neurosci 2001; 24: 1217-81.

[81]   De Cesare D, Jacquot S, Hanauer A, Sassone-Corsi P. Rsk-2 activity is necessary for epidermal growth factor-induced phosphorylation of CREB protein and transcription of c-fos gene. Proc Natl Acad Sci USA 1998; 95: 12202-7.

[82]   Xing J, Ginty DD, Greenberg ME. Coupling of the RAS-MAPK pathway to gene activation by RSK2, a growth factor regulated CREB kinase. Science 1996; 273: 959-63.

[83]   Xing J, Kornhauser JM, Xia Z, Thiele EA, Greenberg ME. Nerve growth factor activates extracellular signal-regulated kinase and p38 mitogen-activated protein kinase pathways to stimulate CREB serine 133 phosphorylation. Mol Cell Biol 1998; 18: 1946-55.

[84]   Arthur JS, Cohen P. MSK1 is required for CREB phosphorylation in response to mitogens in mouse embryonic stem cells. FEBS Lett 2000; 482: 44-8.

[85]   Wiggin GR, Soloaga A, Foster JM, Murray-Tait V, Cohen P, Arthur JS. MSK1 and MSK2 are required for the mitogen- and stress-induced phosphorylation of CREB and ATF1 in fibroblasts. Mol Cell Biol 2002; 22: 2871-81.

[86]   Cantley LC. The phosphoinositide 3-kinase pathway. Science 2002; 296: 1655-7.

[87]   Du K, Montminy M. CREB is a regulatory target for the protein kinase Akt/PKB. J Biol Chem 1998; 273: 32377-9.

[88]   Pugazhenthi S, Nesterova A, Sable C, et al. Akt/protein kinase B up-regulates Bcl-2 expression through cAMP-response element-binding protein. J Biol Chem 2000; 275: 10761-6.

[89]   Lin CH, Yeh SH, Lu KT, Leu TH, Chang WC, Gean PW. A role for the PI-3 kinase signaling pathway in fear conditioning and synaptic plasticity in the amygdala. Neuron 31; 2001: 841-51.

[90]   Perkinton MS, Ip JK, Wood GL, Crossthwaite AJ, Williams RJ. Phosphatidylinositol 3-kinase is a central mediator of NMDA receptor signalling to MAP kinase (Erk1/2), Akt/PKB and CREB in striatal neurones. J Neurochem 2002; 80: 239-54.

[91]   Riccio A, Pierchala BA, Ciarallo CL, Ginty DD. An NGF-TrkA-mediated retrograde signal to transcription factor CREB in sympathetic neurons. Science 1997; 277: 1097-1100.

[92]   Watson FL, Heerssen HM, Moheban DB, et al. Rapid nuclear responses to target-derived neurotrophins require retrograde transport of ligand-receptor complex. J Neurosci 1999; 19: 7889-7900.

[93]   Watson FL, Heerssen HM, Bhattacharyya A, Klesse L, Lin MZ, Segal RA. Neurotrophins use the Erk5 pathway to mediate a retrograde survival response. Nat Neurosci 2001; 4: 981-8.

[94]   Alberts AS, Montminy M, Shenolikar S, Feramisco JR. Expression of a peptide inhibitor of protein phosphatase 1 increases phosphorylation and activity of CREB in NIH 3T3 fibroblasts. Mol Cell Biol 1994; 14: 4398-407.

[95]   Hagiwara M, Alberts A, Brindle P, et al. Transcriptional attenuation following cAMP induction requires PP-1-mediated dephosphorylation of CREB. Cell 1992; 70: 105-13.

[96]   Wadzinski BE, Wheat WH, Jaspers S, et al. Nuclear protein phosphatase 2A dephosphorylates protein kinase A-phosphorylated CREB and regulates CREB transcriptional stimulation. Mol Cell Biol 1993; 13: 2822-34.

[97]   Bito H, Deisseroth K, Tsien RW. CREB phosphorylation and dephosphorylation: a Ca^{2+}- and stimulus duration-dependent switch for hippocampal gene expression. Cell 1996; 87: 1203-14.

[98]   Genoux D, Haditsch U, Knobloch M, Michalon A, Storm D, Mansuy IM. The protein phosphatase 1 is a molecular constraint on learning and memory. Nature 2002; 418: 970975.

[99]   Hardingham GE, Fukunaga Y, Bading H. Extrasynaptic NMDARs oppose synaptic NMDARs by triggering CREB shutoff and cell death pathways. Nat Neurosci 2002; 5: 405-14.

[100]   Sala C, Rudolph-Correia S, Sheng M. Developmentally regulated NMDA receptor-dependent dephosphorylation of cAMP response element-binding protein (CREB) in hippocampal neurons. J. Neurosci 2000; 20: 3529-36.

[101]   Cha-Molstad H, Keller DM, Yochum GS, Impey S, Goodman,RH. Cell-type-specific binding of the transcription factor CREB to the cAMP-response element. Proc Natl Acad Sci USA 2004; 101: 13572-7.

[102]   Conkright MD, Guzman E, Flechner L, Su AI, Hogenesch JB, Montminy M. Genome-wide analysis of CREB target genes reveals a core promoter requirement for cAMP responsiveness. Mol Cell 2003; 11: 1101-8.

[103]   Zhang X, Odom DT, Koo SH, et al. Genome-wide analysis of cAMP-response element binding protein occupancy, phosphorylation, and target gene activation in human tissues. Proc Natl Acad Sci USA 2005; 102: 4459-64.

[104]   Deisseroth K, Tsien RW. Dynamic multiphosphorylation passwords for activity-dependent gene expression. Neuron 2002; 34: 179-82.

[105]   Goodman RH, Smolik S. CBP/p300 in cell growth, transformation, development. Genes Dev 2000; 14: 1553-77.

[106]   Kwok RPS, Lundblad JR, Chrivia JC, et al. Nuclear protein CBP is a coactivator for the transcription factor CREB. Nature 1994; 370: 223-6.

[107]   Hu SC, Chrivia J, Ghosh A. Regulation of CBP-mediated transcription by neuronal calcium signaling. Neuron 1999; 22: 799-808.

[108]   Impey S, Fong AL, Wang Y, et al. Phosphorylation of CBP mediates transcriptional activation by neural activity and CaM kinase IV. Neuron 2002; 34: 235-44.

[109]   Xing L, Quinn PG. Three distinct regions within the constitutive activation domain of cAMP regulatory element-binding protein (CREB) are required for transcription activation. J Biol Chem 1994; 269: 28732-6.

[110]   Felinski EA, Kim J, Lu, Quinn PG. Recruitment of an RNA polymerase II complex is mediated by the constitutive activation domain in CREB, independently of CREB phosphorylation. Mol Cell Biol 2001; 21: 1001-10.

[111]   Felinski EA, Quinn PG. The CREB constitutive activation domain interacts with TATA-binding protein-associated factor 110 (TAF110) through specific hydrophobic residues in one of the three subdomains required for both activation and TAF110 binding. J Biol Chem 1999; 274: 11672-8.

[112]   Nakajima T, Uchida C, Anderson SF, et al. RNA helicase A mediates association of CBP with RNA polymerase II. Cell 1997; 90: 1107-12.

[113]   Swope DL, Mueller CL, Chrivia JC. CREB-binding protein activates transcription through multiple domains. J Biol Chem 1996; 271: 28138-45.

[114]   Bannister AJ, Kouzarides T. The CBP co-activator is a histone acetyltransferase. Nature 1996; 384: 641-3.

[115]   Liu FC, Graybiel AM. Spatiotemporal dynamics of CREB phosphorylation: transient versus sustained phosphorylation in the developing striatum. Neuron 1996; 17: 1133-44.

[116]   Mayr BM, Canettieri G, Montminy MR. Distinct effects of cAMP and mitogenic signals on CREB-binding protein recruitment impart specificity to target gene activation via CREB. Proc Natl Acad Sci USA 2001; 98: 10936-41.

[117]   Thompson MA, Ginty DD, Bonni A, Greenberg ME. L-type voltage-sensitive Ca^{2+} channel activation regulates c-fos transcription at multiple levels. J Biol Chem 1995; 270: 4224-35.

[118]   Parker D, Jhala US, Radhakrishnan I, et al. Analysis of an activator:coactivator complex reveals an essential role for secondary structure in transcriptional activation. Mol Cell 1998; 2: 353-9.

[119]   Gau D, Lemberger T, von Gall C, et al. Phosphorylation of CREB Ser142 regulates light-induced phase shifts of the circadian clock. Neuron 2002; 34: 245-53.

[120]   Kornhauser JM, Cowan CW, Shaywitz AJ, *et al*. CREB transcriptional activity in neurons is regulated by multiple, calcium-specific phosphorylation events. Neuron 2002; 34: 221-33.

[121]   Iourgenko V, Zhan W, Mickanin C, *et al*. Identification of a family of cAMP response element-binding protein coactivators by genome-scale functional analysis in mammalian cells. Proc Natl Acad Sci USA 2003; 100: 12147-52.

[122]   Bittinger MA, McWhinnie E, Meltzer J, *et al*. Activation of cAMP response element-mediated gene expression by regulated nuclear transport of TORC proteins. Curr Biol 2004; 14: 2156-61.

[123]   Koo SH, Flechner L, Qi L, *et al*. The CREB coactivator TORC2 is a key regulator of fasting glucose metabolism. Nature 2005; 37: 1109-11.

[124]   Screaton RA, Conkright MD, Katoh Y, *et al*. The CREB coactivator TORC2 functions as a calcium- and cAMP-sensitive coincidence detector. Cell 2004; 119: 61-74.

[125]   Xu W, Kasper LH, Lerach S, Jeevan T, Brindle P.K. Individual CREB-target genes dictate usage of distinct cAMP-responsive coactivation mechanisms. EMBO J 2007; 26: 2890-2903.

[126]   Ravnskjaer K, Kester H, Liu Y, *et al*. Cooperative interactions between CBP and TORC2 confer selectivity to CREB target gene expression. EMBO J 2007; 26: 2880-9.

[127]   Carrión AM, Link WA, Ledo F, Mellstrom B, Naranjo JR. DREAM is a Ca^{2+}-regulated transcriptional repressor. Nature 1999; 398: 80-4.

[128]   Ledo F, Kremer L, Mellström B, Naranjo JR. Ca2+-dependent block of CREB-CBP transcription by repressor DREAM. EMBO J 2002; 21: 4583-92.

[129]   Ledo F, Carrión AM, Link WA, Mellström B, Naranjo JR. DREAM-alphaCREM interaction *via* leucine-charged domains derepresses downstream regulatory element-dependent transcription. Mol Cell Biol 2000; 20: 9120-6.

[130]   Sassone-Corsi P. Transcription factors responsive to cAMP. Annu Rev Cell Dev Biol 1995; 11: 355-77.

[131]   Stehle JH, Foulkes NS, Molina CA, Simonneaux V, Pevet P, Sassone-Corsi P. Adrenergic signals direct rhythmic expression of transcriptional repressor CREM in the pineal gland. Nature 1993; 365: 314-20.

[132]   Molina CA, Foulkes NS, Lalli E, Sassone-Corsi P. Inducibility and negative autoregulation of CREM: an alternative promoter directs the expression of ICER, an early response repressor. Cell 1993; 75: 875-86.

[133]   Hu Y, Lund IV, Gravielle MC, Farb DH, Brooks-Kayal AR, Russek SJ. Surface expression of GABAA receptors is transcriptionally controlled by the interplay of cAMP-response element-binding protein and its binding partner inducible cAMP early repressor. J Biol Chem 2008; 283: 9328-40.

[134]   Michael LF, Asahara H, Shulman AI, Kraus WL, Montminy M. The phosphorylation status of a cyclic AMP-responsive activator is modulated *via* a chromatin-dependent mechanism. Mol Cell Biol 2000; 20: 1596-603.

[135]   Canettieri G, Morantte I, Guzmán E, *et al*. Attenuation of a phosphorylation-dependent activator by an HDAC-PP1 complex. Nat Struct Biol 2003; 10: 175-81.

[136]   Gao J, Siddoway B, Huang Q, Xia H. Inactivation of CREB mediated gene transcription by HDAC8 bound protein phosphatase. Biochem Biophys Res Commun 2009; 379: 1-5.

[137]   Guan Z, Giustetto M, Lomvardas S, *et al*. Integration of long-term-memory-related synaptic plasticity involves bidirectional regulation of gene expression and chromatin structure. Cell 2002; 111: 483-93.

[138]   Impey S, McCorkle SR, Cha-Molstad H, *et al*. Defining the CREB regulon: a genome-wide analysis of transcription factor regulator regions. Cell 2004; 119: 1041-54.

[139]   McClung CA, Nestler EJ. Regulation of gene expression and cocaine reward by CREB and DFosB. Nat Neurosci 2003; 6: 1208-15.

[140]   Lane-Ladd, SB, Pineda J, Boundy VA, *et al*. CREB (cAMP response element-binding protein) in the locus coeruleus: biochemical, physiological, and behavioral evidence for a role in opiate dependence. J Neurosci 1997; 17: 7890-901.

[141]   Kim KS, Lee MK, Carroll J, Joh TH. Both the basal and inducible transcription of the tyrosine hydroxylase gene are dependent upon a cAMP response element. J Biol Chem 1993; 268: 15689-95.

[142]   Itoi K, Horiba N, Tozawa F, *et al*. Major role of 3′, 5′-cyclic adenosine monophosphate dependent protein kinase A pathway in corticotropin-releasing factor gene expression in the rat hypothalamus in vivo. Endocrinology 1996; 137: 2389-96.

[143]   Douglass J, McKinzie AA, Pollock KM. Identification of multiple DNA elements regulating basal and protein kinase A-induced transcriptional expression of the rat prodynorphin gene. Mol. Endocrinol 1994; 8: 333-44.

[144]   Cole RL, Konradi C, Douglass J, Hyman SE. Neuronal adaptation to amphetamine and dopamine: molecular mechanisms of prodynorphin gene regulation in rat striatum. Neuron 1995; 14: 813-23.

[145]   Carlezon WA Jr, Thome J, Olson VG, *et al*. Regulation of cocaine reward by CREB. Science 1998; 282: 2272-5.

[146]   Sakai N, Thome J, Newton SS, *et al*. Inducible and brain region specific CREB transgenic mice. Mol Pharmacol 2002; 61: 1453-64.

[147]   Borges K, Dingledine R. Functional organization of the GluR1 glutamate receptor promoter. J Biol Chem 2001; 276: 25929-38.

[148]   Olson VG, Zabetian CP, Bolanos CA, *et al*. Regulation of drug reward by CREB: Evidence for two functionally distinct subregions of the ventral tegmental area. J Neurosci 2005; 25: 5553-62.

[149]   Kandel ER. The molecular biology of memory storage: a dialogue between genes and synapses. Science 2001; 294: 1030-8.

[150]   Silva AJ, Kogan JH, Frankland PW, Kida S. CREB and memory. Annu Rev Neurosci 1998; 21: 127-48.

[151]   Montarolo PG, Goelet P, Castellucci VF, Morgan J, Kandel ER, Schacher SA. A critical period for macromolecular synthesis in long-term heterosynaptic facilitation in Aplysia. Science 1986; 234: 1249-54.

[152]   Kandel ER. Genes, nerve cells, and the remembrance of things past. .J Neuropsychiatry Clin. Neurosci 1989; 1: 103-25.

[153]   Martin KC, Michael D, Rose JC, *et al*. MAP kinase translocates into the nucleus of the presynaptic cell and is required for long-term facilitation in *Aplysia*. Neuron 1997; 18: 899-912.

[154]   Michael D, Martin KC, Seger R, Ning MM, Baston R, Kandel ER. Repeated pulses of serotonin required for long-term facilitation activate mitogen-activated protein kinase in sensory neurons of *Aplysia*. Proc Natl Acad Sci USA 1998; 95: 1864-9.

[155]   Dash PK, Hochner B, Kandel ER. Injection of the cAMP-responsive element into the nucleus of Aplysia sensory neurons blocks long-term facilitation. Nature 1990; 345: 718-21.

[156]   Alberini CM, Ghirardi M, Metz R, Kandel ER. C/EBP is an immediate-early gene required for the consolidation of long-term facilitation in Aplysia. Cell 1994; 76: 1099-114.

[157]   Bartsch D, Ghirardi M, Skehel PA, *et al*. Aplysia CREB2 represses long-term facilitation: relief of repression converts transient facilitation into long-term functional and structural change. Cell 1995; 83: 979-92.

[158]   Bartsch D, Casadio A, Karl K.A, Serodio P, Kandel ER. CREB1 encodes a nuclear activator, a repressor, and a cytoplasmic modulator that form a regulatory unit critical for long-term facilitation. Cell 1998, 95, 211-223.

[159] Kaang BK, Kandel ER, Grant SG. Activation of cAMP-responsive genes by stimuli that produce long-term facilitation in Aplysia sensory neurons. Neuron 1993; 10: 427-35.

[160] Waddell S, Quinn WG. Flies, genes, and learning. Annu Rev Neurosci 2001; 24: 1283-309.

[161] Yin JC, Wallach JS, Del Vecchio M, et al. Induction of a dominant negative CREB transgene specifically blocks long-term memory in Drosophila. Cell 1994; 79: 49-58.

[162] Yin JC, Del Vecchio M, Zhou H, Tully,T. CREB as a memory modulator: induced expression of a dCREB2 activator isoform enhances long-term memory in Drosophila. Cell 1995; 81: 107-15.

[163] Wu ZL, Thomas SA, Villacres EC, Xia Z, et al. Altered behavior and long-term potentiation in type I adenylyl cyclase mutant mice. Proc Natl Acad Sci USA 1995; 92: 220-4.

[164] Huang YY, Li XC, Kandel ER. cAMP contributes to mossy fiber LTP by initiating both a covalently mediated early phase and macromolecular synthesis-dependent late phase. Cell 1994; 79: 69-79.

[165] Frey U, Huang YY, Kandel ER. Effects of cAMP simulate a late stage of LTP in hippocampal CA1 neurons. Science 1993; 260: 1661-4.

[166] Abel T, Nguyen PV, Barad M, Deuel TA, Kandel ER, Bourtchouladze R. Genetic demonstration of a role for PKA in the late phase of LTP and in hippocampus-based long-term memory. Cell 1997; 88: 615-26.

[167] Pham TA, Impey S, Storm DR, Stryker MP. CRE-mediated gene transcription in neocortical neuronal plasticity during the developmental critical period. Neuron 1999; 22: 63-72.

[168] Deisseroth K, Bito H, Tsien RW. Signaling from synapse to nucleus: postsynaptic CREB phosphorylation during multiple forms of hippocampal synaptic plasticity. Neuron 1996; 16: 89-101.

[169] Impey S, Mark M, Villacres EC, Poser S, Chavkin C, Storm DR. Induction of CRE-mediated gene expression by stimuli that generate long-lasting LTP in area CA1 of the hippocampus. Neuron 1996; 16: 973-82.

[170] Impey S, Smith D.M, Obrietan K, Donahue R, Wade C, Storm DR. Stimulation of cAMP response element (CRE)-mediated transcription during contextual learning. Nat. Neurosci 1998; 1: 595-601.

[171] Schulz S, Siemer H, Krug M, Hollt V. Direct evidence for biphasic cAMP responsive element-binding protein phosphorylation during long-term potentiation in the rat dentate gyrus in vivo. J Neurosci 1999; 19: 5683-92.

[172] Stanciu M, Radulovic J, Spiess J. Phosphorylated cAMP response element binding protein in the mouse brain after fear conditioning: relationship to Fos production. Brain Res Mol Brain Res 2001; 94: 15-24.

[173] Taubenfeld SM, Wiig KA, Monti B, Dolan B, Pollonini G, Alberini CM. Fornix-dependent induction of hippocampal CCAAT enhancer-binding protein β and Λ co-localizes with cAMP response element-binding protein and accompanies long-term memory consolidation. J Neurosci 2001; 21: 84-91.

[174] Viola H, Furman M, Izquierdo LA, et al. Phosphorylated cAMP response element-binding protein as a molecular marker of memory processing in rat hippocampus: effect of novelty. J Neurosci 2000; 20: RC112.

[175] Guzowski JF, McGaugh JL. Antisense oligodeoxy-nucleotide-mediated disruption of hippocampal cAMP response element binding protein levels impairs consolidation of memory for water maze training. Proc Natl Acad Sci USA 1997; 94: 2693-8.

[176] Ahn S, Ginty DD, Linden DJ. A late phase of cerebellar long-term depression requires activation of CaMKIV and CREB. Neuron 1999; 23: 559-68.

[177] Hummler E, Cole TJ, Blendy JA, et al. Targeted mutation of the CREB gene: compensation within the CREB/ATF family of transcription factors. Proc Natl Acad Sci USA 1994; 91: 5647-51.

[178] Blendy JA, Kaestner KH, Schmid W, Gass P, Schutz G. Targeting of the CREB gene leads to up regulation of a novel CREB mRNA isoform. EMBO J 1996; 15: 1098-106.

[179] Bourtchuladze R, Frenguelli B, Blendy J, Cioffi D, Schutz G, Silva AJ. Deficient long-term memory in mice with a targeted mutation of the cAMP-responsive element-binding protein. Cell, 1994; 79: 59-68.

[180] Gass P, Wolfer DP, Balschun D, et al. Deficits in memory tasks of mice with CREB mutations depend on gene dosage. Learn. Mem 1998; 5: 274-88.

[181] Kogan JH, Frankland PW, Blendy JA, et al. Spaced training induces normal long-term memory in CREB mutant mice. Curr Biol 1997; 7: 1-11.

[182] Pittenger C, Huang YY, Paletzki RF, et al. Reversible inhibition of CREB/ATF transcription factors in region CA1 of the dorsal hippocampus disrupts hippocampus-dependent spatial memory. Neuron 2002; 34: 447-62.

[183] Barco A, Alarcón JM, Kandel ER. Expression of constitutively active CREB protein facilitates the late phase of long- term potentiation by enhancing synaptic capture. Cell 2002; 108: 689-703.

[184] Viosca J, López de Armentia M, Jancic D, Barco A. Enhanced CREB-dependent gene expression increases the excitability of neurons in the basal amygdala and primes the consolidation of contextual and cued fear memory. Learn Mem 2009a; 16: 193-7.

[185] Lopez de Armentia M, Jancic D, Olivares R, Alarcón JM, Kandel ER, Barco A. cAMP response element-binding protein-mediated gene expression increases the intrinsic excitability of CA1 pyramidal neurons. J Neurosci 2007; 27: 13909-18.

[186] Viosca J, Malleret G, Bourtchouladze R, et al. Chronic enhancement of CREB activity in the hippocampus interferes with the retrieval of spatial information. Learn Mem 2009b; 16: 198-209.

[187] Kida S, Josselyn SA, de Ortiz SP, et al. CREB required for the stability of new and reactivated fear memories. Nat Neurosci 2002; 5: 348-55.

[188] Jancic D, López de Armentia M, Valor LM, Olivares R, Barco A. Inhibition of cAMP Response Element-Binding Protein Reduces Neuronal Excitability and Plasticity, and Triggers Neurodegeneration. Cerebral Cortex 2009; doi:10.1093/cercor/bhp004

[189] Fontán-Lozano A, Romero-Granados R, del-Pozo-Martín Y, et al. Lack of DREAM protein enhances learning and memory and slows brain aging. Curr Biol 2009; 19: 54-60.

[190] Alexander JC, McDermott CM, Tunur T, et al. The role of calsenilin/DREAM/KChIP3 in contextual fear conditioning. Learn Mem 2009; 16: 167-77.

[191] Kojima N, Borlikova G, Sakamoto T, et al. Inducible cAMP early repressor acts as a negative regulator for kindling epileptogenesis and long-term fear memory. J Neurosci 2008; 28: 6459-72.

[192] Kovacs KA, Steullet P, Steinmann M, et al. TORC1 is a calcium- and cAMP-sensitive coincidence detector involved in hippocampal long-term synaptic plasticity. Proc Natl Acad Sci USA 2007; 104: 4700-5.

[193] Zhou Y, Wu H, Li S, et al. Requirement of TORC1 for late-phase long-term potentiation in the hippocampus. PLoS One 2006; 1: E16.

[194] Wu H, Zhou Y, Xiong ZQ. Transducer of regulated CREB and late phase long-term synaptic potentiation. FEBS J 2007; 274: 3218-23.

[195] Josselyn SA. What's right with my mouse model? New insights into the molecular and cellular basis of cognition from mouse models of Rubinstein-Taybi Syndrome. Learn Mem 2005; 12: 80-3.

[196] Waddell S. Protein phosphatase 1 and memory: practice makes PP1 imperfect? Trends Neurosci 2003; 26: 117-9.

[197]   Guan JS, Haggarty SJ, Giacometti E, *et al.* HDAC2 negatively regulates memory formation and synaptic plasticity. Nature 2009; 459: 55-60.

[198]   Bernabeu R, Bevilaqua L, Ardenghi P, *et al.* Involvement of hippocampal cAMP/cAMP-dependent protein kinase signaling pathways in a late memory consolidation phase of aversively motivated learning in rats. Proc Natl Acad Sci USA 1997; 94: 7041-6.

[199]   Zhao W, Bennett P, Sedman GL. The impairment of long-term memory formation by the phosphatase inhibitor okadaic acid. Brain Res Bull 1995; 36: 557-61.

[200]   Ahmed T, Frey JU. Plasticity-specific phosphorylation of CaMKII, MAP-kinases and CREB during late-LTP in rat hippocampal slices in vitro. Neuropharmacology 2005; 49: 477-92.

[201]   Leutgeb JK, Frey JU, Behnisch T. Single cell analysis of activity dependent cyclic AMP-responsive element-binding protein phosphorylation during long-lasting long-term potentiation in area CA1 of mature rat hippocampal-organotypic cultures. Neuroscience 2005; 131: 601-10.

[202]   Han JH, Kushner SA, Yiu AP, *et al.* Neuronal competition and selection during memory formation. Science 2007; 316: 457-60.

[203]   Casadio A, Martin KC, Giustetto M, *et al.* A transient, neuron-wide form of CREB-mediated long-term facilitation can be stabilized at specific synapses by local protein synthesis. Cell 1999; 99: 221-37.

[204]   Martin KC, Casadio A, Zhu H, EY, *et al.* Synapse-specific, long-term facilitation of aplysia sensory to motor synapses: a function for local protein synthesis in memory storage. Cell 1997; 91: 927-38.

[205]   Frey U, Morris RG. Synaptic tagging and long-term potentiation. Nature 1997; 385: 533-6.

[206]   Carlezon WA Jr, Duman RS, Nestler EJ. The many faces of CREB. Trends Neurosci 2005; 28: 436-45.

[207]   McClung CA, Nestler EJ. Regulation of gene expression and cocaine reward by CREB and DFosB. Neuropsychopharmacology 2008; 33: 3-17.

[208]   Duman RS. Pathophysiology of depression: the concept of synaptic plasticity. Eur Psychiatry 2002; 17(Suppl 3): 306-10.

[209]   Blendy JA. The role of CREB in depression and antidepressant treatment. Biol Psychiatry 2006; 59: 1144-50.

[210]   Nestler EJ, Carlezon Jr WA. The mesolimbic dopamine reward circuit in depression. Biol Psychiatry 2006; 59: 1151-9.

[211]   Chen AC, Shirayama Y, Shin KH, Neve RL, Duman RS. Expression of the cAMP response element binding protein (CREB) in hippocampus produces an antidepressant effect. Biol Psychiatry 2001a; 49: 753-62.

[212]   Pliakas AM, Carlson RR, Neve RL, Konradi C, Nestler EJ, Carlezon Jr WA. Altered responsiveness to cocaine and increased immobility in the forced swim test associated with elevated cAMP response element-binding protein expression in nucleus accumbens. J Neurosci 2001; 21: 7397-403.

[213]   Newton SS, Thome J, Wallace TL, *et al.* Inhibition of cAMP response element binding protein or dynorphin in the nucleus accumbens produces an antidepressant-like effect. J Neurosci 2002; 22: 10883-90.

[214]   Wallace TL, Stellitano KE, Neve RL, Duman RS. Effects of cyclic adenosine monophosphate response element binding protein overexpression in the basolateral amygdala on behavioral models of depression and anxiety. Biol Psychiatry 2004; 56: 151-60.

[215]   Duman RS, Heninger GR, Nestler EJ. A molecular and cellular hypothesis of depression. Arch Gen Psychiatry 1997; 54: 597-606.

[216]   Newton SS, Collier EF, Hunsberger J, *et al.* Gene profile of electroconvulsive seizures: induction of neurotrophic and angiogenic factors. J Neurosci 2003; 23: 10841-51.

[217]   Nibuya M, Morinobu S, Duman RS. Regulation of BDNF and trkB mRNA in rat brain by chronic electroconvulsive seizure and antidepressant drug treatments. J Neurosci 1995; 15: 7539-47.

[218]   Duman RS. Depression: a case of neuronal life and death? Biol Psychiatry 2004; 56: 140-5.

[219]   Nestler EJ, Barrot M, DiLeone RJ, Eisch AJ, Gold SJ, Monteggia LM. Neurobiology of depression. Neuron 2002; 34: 13-25

[220]   Todtenkopf MS, Marcus JF, Portoghese PS, Carlezon WA Jr. Effects of k-opioid receptor ligands on intracranial self-stimulation in rats. Psychopharmacology 2004; 172: 463-70.

[221]   McLaughlin JP, Marton-Popovici M, Chavkin C. Kappa opioid receptor antagonism and prodynorphin gene disruption block stress-induced behavioural responses. J Neurosci 2003; 23: 5674-83.

[222]   Mague SD, Pliakas AM, Todtenkopf MS, *et al.* Antidepressant-like effects of k-opioid receptor antagonists in the forced swim test in rats. J Pharmacol Exp Ther 2003; 305: 323-30.

[223]   Di Chiara G, Imperato A. Drugs abused by humans preferentially increase synaptic dopamine concentrations in the mesolimbic system of freely moving rats. Proc Natl Acad Sci USA 1988; 85: 5274-8.

[224]   Shippenberg TS, Rea W. Sensitization to the behavioural effects of cocaine: modulation by dynorphin and k-opioid receptor agonists. Pharmacol Biochem Behav 1997; 57: 449-55.

[225]   Svingos AL, Moriwaki A, Wang JB, Uhl GR, Pickel VM. Cellular sites for dynorphin activation of k-opioid receptors in the rat nucleus accumbens shell. J Neurosci 1999; 19: 1804-1813.

[226]   Heimer L, Alheid GF, de Olmos JS, *et al.* The accumbens: beyond the core-shell dichotomy. J Neuropsychiatry Clin Neurosci 1997; 9: 354-81.

[227]   Barrot, M, Olivier JD, Perrotti LI, *et al.* CREB activity in the nucleus accumbens shell controls gating of behavioral responses to emotional stimuli. Proc Natl Acad Sci USA 2002; 99: 11435-40.

[228]   Barrot M, Wallace DL, Bolaños CA, *et al.* Regulation of anxiety and initiation of sexual behavior by CREB in the nucleus accumbens. Proc Natl Acad Sci USA 2005; 102: 8357-62

[229]   Pandey SC, Roy A, Zhang H. The decreased phosphorylation of cyclic adenosine monophosphate (cAMP) response element binding (CREB) protein in the central amygdala acts as a molecular substrate for anxiety related to ethanol withdrawal in rats. Alcohol Clin Exp Res 2003; 27: 396-409.

[230]   Valverde O, Mantamadiotis T, Torrecilla M, *et al.* Modulation of anxiety-like behavior and morphine dependence in CREB-deficient mice. Neuropsychopharmacol 2004; 29: 1122-33.

[231]   Barco A, Pittenger C, Kandel ER. CREB, memory enhancement and the treatment of memory disorders: promises, pitfalls and prospects Expert Opin Ther Targets 2003; 7: 101-14.

[232]   Tanis KQ, Duman RS, Newton SS. CREB binding and activity in brain: regional specificity and induction by electroconvulsive seizure Biol Psychiatry 2008; 63: 710-20.

[233]   Lonze BE, Ginty DD. Function and regulation of CREB family transcription factors in the nervous system Neuron 2002; 35: 605-23.

CHAPTER 3

Transcriptional Profiling of Hippocampal Memory-Associated Synaptic Plasticity: Old Friends and New Faces

Niamh C. O'Sullivan, Graham K. Sheridan and Keith J. Murphy[*]

Applied Neurotherapeutics Research Group, UCD School of Biomolecular and Biomedical Science, Conway Institute, University College Dublin, Belfield, Dublin 4, Ireland

Abstract: Information storage is a fundamental capacity of neuronal circuits that underpins all higher cognitive functions including long-term memory formation, working memory, behavioural control and language. The storage of information requires alterations in strength and pattern of synaptic connections in key brain structures such as the hippocampus. It is now clear that such memory-associated synaptic plasticity is driven by a cascade of gene transcription and new protein synthesis. Here, we review how the use of high-throughput microarray platforms and bioinformatic *in silico* analyses is now revealing an extensive, integrated transcriptional programme underpinning synaptic plasticity that confirms roles for the previously well-characterised transcription factors NF-κB and CREB but also implicates more novel players such as SRF, NFAT and HIF-1. The transcriptional programme likely sees recruitment of tens of transcription factors and hundreds of genes, orchestrated through the three core periods of synapse destabilisation, new synapse construction and selective synapse retention. We discuss the nature of the contributions of NF-κB, CREB, SRF and NFAT to cognition-associated synaptic plasticity and present new data to support a biphasic role of HIF-1 during the early memory consolidation period.

Keywords: Water maze, passive avoidance, hippocampus, long-term memory, Transcription control, HIF-1, SRF, NFAT, CREB, NF-kB, microarray, transcription factor motif.

INTRODUCTION

The remarkable capacity of the central nervous system to learn and retain new information is at the very heart of higher cognitive functions exemplified by long-term memory, working memory, language and behavioural control. It is now well accepted that information storage can be divided into short and long-lasting forms. Short-term memory encompasses the initial acquisition phase of a particular memory and also the construct of working memory. It usually lasts seconds to minutes and can occur in numerous brain structures including hippocampus and prefrontal cortex. Short-term memory is predominantly accomplished through post-translational and/or sub-cellular shuttling of pre-existing molecules, such as receptors and ion-channels, which can rapidly alter the efficiency of synaptic transmission [1,2]. In contrast, long-term memory requires consolidation and can persist for days, weeks or even years.

For over a century now the fundamental nature of information storage in the brain has been hypothesised to lie in an ability to alter synaptic connectivity within key circuitries such as the hippocampus and neocortex [3-5]. The drive to understand synaptic plasticity has seen the convergence of several fields including electrophysiology, cognitive neuroscience, developmental biology and molecular biology to investigate the fundamental underlying molecular and cellular processes that allows such remarkable, controlled reorganisation in brain circuitry. Synaptic plasticity and reorganisation of neuronal connections is best studied within the hippocampus where it is vital for the formation

*Address correspondence to Keith J. Murphy: Applied Neurotherapeutics Research Group, UCD School of Biomolecular and Biomedical Science, Conway Institute, University College Dublin, Belfield, Dublin 4, Ireland; Tel: +353-1-7166778; Fax: +353-1-7166920; E-mail: Keith.Murphy@ucd.ie

and consolidation of new memories [6,7]. Although there has been much debate as to the nature of such hippocampal plasticity, it is now clear that both functional (synapse strength) and morphological (synapse size and position) plasticity occur [8-10].

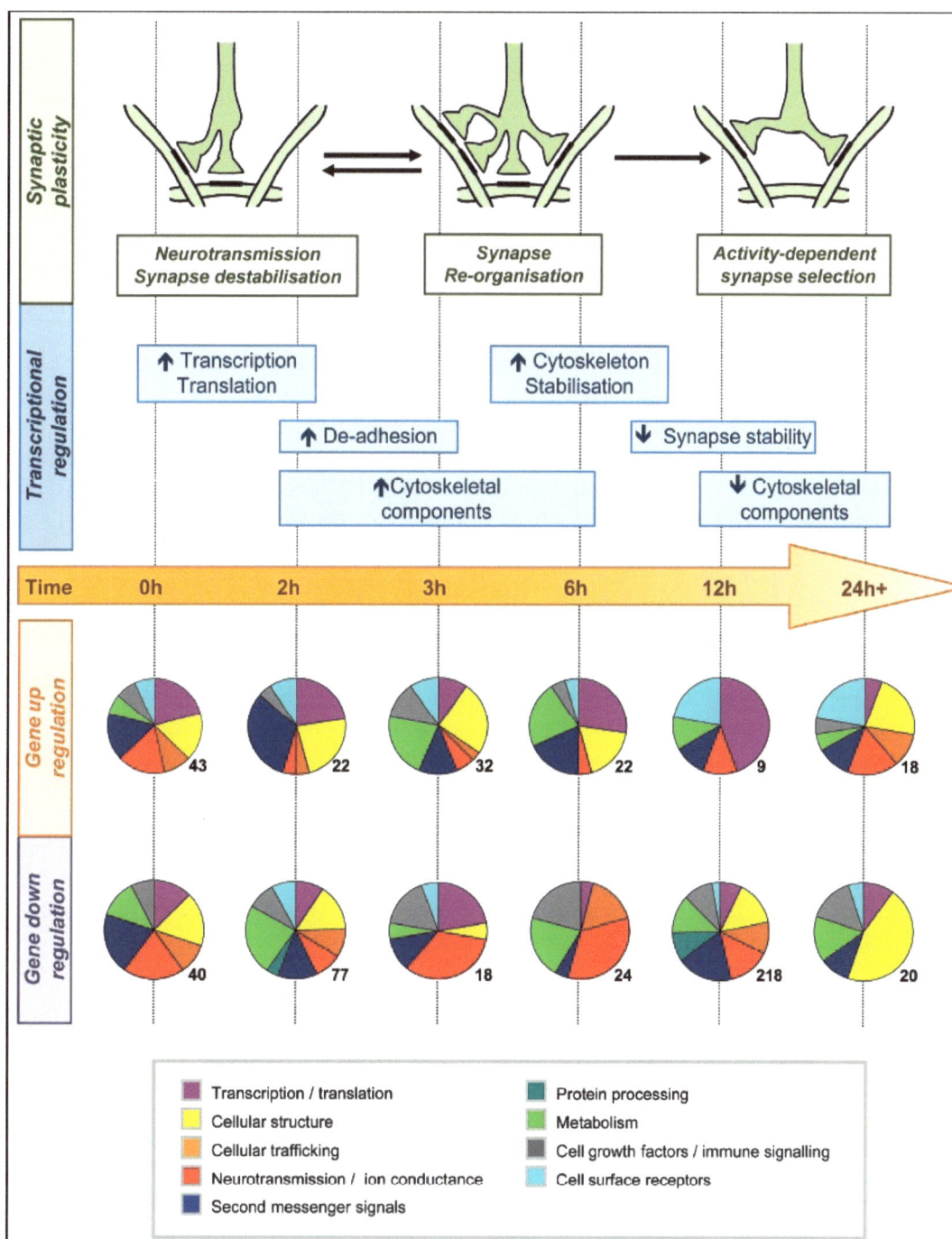

Fig. (1). Temporal profile of transcriptional events in the dentate gyrus over the 24 hours post-training period. The top panel highlights the timing of the key phases of memory-associated synaptic plasticity initiated by a learning event in the rodent hippocampal dentate gyrus. The next panel summarises the major functional regulations implicated by transcriptional profiling studies of memory consolidation in the hippocampus. The arrow in each box indicates whether this functional group is up or down regulated at the transcriptional level. The bottom two panels show the functional group breakdown of genes regulating at each time point post-passive avoidance learning based on the microarray study by O'Sullivan *et al.* [35]. The number beside each pie chart indicates the total number of genes represented.

Memory-associated hippocampal synaptic plasticity is, by its very nature, intertwined with the cardinal aspect of time. Fundamentally, memory consolidation should be viewed as a temporal cascade of directed synaptic change tightly controlled in time and space to produce a reorganised neuronal connectivity. In the hippocampus it is helpful to divide synaptic plasticity into three key phases initiated by a learning event: synaptic destabilisation, synaptic reorganisation and synaptic selection (Fig. 1). Synaptic destabilisation is essentially a preparative stage and is marked by events such as selective internalisation and degradation of adhesive elements, such as neural cell adhesion molecule (NCAM) and amyloid precursor protein (APP) and remodeling of the extracellular matrix through regulation of perisynaptic protease activity [11-13]. During synaptic reorganisation changes in synaptic number, synaptic curvature, size of synaptic elements and synaptic perforations have been observed [14,15]. For example, quantitative electron microscopic analysis of the rat hippocampal dentate gyrus has revealed an increase in dendritic spine and synaptic density, 6-9h following avoidance conditioning and spatial learning [16-18]. This increase in synaptic number is transient, however, returning to basal levels by 24-48h post-training. Thus, the last phase marks a period of synaptic selection, encompassing the two competing, parallel processes of synapse elimination and stabilization necessary for retention of only relevant connections in the memory circuit [15].

Many studies have employed inhibitors of transcription and translation to show that memory-associated hippocampal synaptic plasticity is driven by a cascade of *de novo* gene transcription and protein synthesis [7,19-23]. Given the complexity of the mammalian nervous system, the elucidation of molecular mechanisms that mediate memory-associated synaptic reorganisation has relied heavily on simple models of learning and synaptic plasticity. For example, the electrophysiological study of long-term potentiation (LTP) and long-term depression (LTD), long-lasting enhancement and decrement in synaptic strength, respectively, are widely believed to model the cellular events occurring during memory consolidation [8]. In addition, animal model organisms such as *Aplysia* and *Drosophila* have allowed elucidation of individual components of the transcriptional programme. The seminal work of Eric Kandel and others on CREB provided the first and best characterised transcriptional regulator with a clear and critical role in cognition-associated synaptic plasticity [24,25]. Such studies implicated transmission-mediated control of cAMP signalling, Ca^{2+}/calmodulin-dependent protein kinase IV (CaMKIV) and protein kinase A in the up-stream regulation of the CREB transcription factor and heralded a rapid expansion in our understanding of how neuronal activity patterns could initiate and control physical synaptic reorganization to lay down new representations [20,26,27].

The advent of high-throughput microarray platforms and more sophisticated bioinformatic *in silico* analyses now leave us in no doubt that CREB is but one component of an extensive, integrated transcriptional programme underpinning synaptic plasticity. This programme likely sees recruitment of tens of transcription factors and hundreds of genes and is carefully orchestrated through the three core periods of synapse destabilisation, new synapse construction and selective synapse retention.

TRANSCRIPTIONAL PROFILING OF MEMORY-ASSOCIATED SYNAPTIC PLASTICITY: COHERENT GENE CLUSTERS AND NOVEL TRANSCRIPTIONAL CONTROL EMERGE

While a range of *in vivo* models of long-term memory have been employed to identify specific transcription factors that regulate gene expression required for consolidation [25,28,29], until recently, the identity and magnitude of gene regulation involved remained poorly characterised. High-throughput methods such as high-density microarray analysis of gene expression have now revealed that large transcriptional programmes are mobilised following learning events such as eye-blink conditioning [30], spatial conditioning [31,32], classical fear conditioning [33,34], and passive avoidance [35]. Below, we briefly discuss the significant challenges that face the application of microarray approaches to the question of the mammalian memory-associated transcriptome. We go on to discuss the function of gene clusters that have emerged, placing them into their temporal context and review the known biology of some of the major transcription factors associated with their regulation during memory consolidation. In addition, we introduce several candidates for the title 'memory-associated transcription factor' that we believe merit further study.

Challenges for Microarray-Based Study of Memory-Associated Synaptic Plasticity

In the first instance, while the application of microarray technology to identify the mammalian synaptic plasticity transcriptome promises much, it is important to recognize that several significant challenges come into play.

1   Dilution effect: To date, microarray approaches have been very successfully applied to study disease state-transcriptomes where gene expression differences can be expected to be substantial and to occur in most cells of the diseased tissue sample employed. Indeed, microarray studies have been informative even in brain tissue in studies of various diseases including multiple sclerosis, Alzheimer's disease and schizophrenia [36-39]. In the setting of memory-associated plasticity in the hippocampus, however, we are dealing with a normal physiological event that likely only recruits 5-10% of the neurons. For example, measurement of the immediate early genes c-fos, Arc and zif-268 following various memory tasks indicates only sparse neuronal recruitment in the hippocampus [32,40]. This introduces a substantial dilution effect that can easily mask a real expression change and certainly reduces fold change values.

2   Cell type heterogeneity: At the cellular level, brain structures are very heterogeneous with neurons, astrocytes, microglia, endothelial cells, and blood cells present in most tissue dissections. Thus, altered mRNA signals likely only occurring in one of these cell types will again be diluted and require follow-up to identify the cell type(s) involved.

3   Transcriptional complexity: The nervous system clearly exhibits the greatest complexity of gene expression of any tissue with upwards of 50–60% of known genes in the human genome expressed in brain. Moreover, many genes are preferentially expressed in the brain [41-43].

4   Given the phases of memory-associated synaptic plasticity outlined above, it is clear that the timing of analysis will have a huge bearing on the nature and function of the transcriptome changes identified. Various studies have focused on distinct temporal periods with one study by O'Sullivan *et al.* [35] attempting a broader temporal characterization. Researchers need to be mindful of what component of memory-associated synaptic plasticity they wish to study and time their post-training sample collection accordingly.

5   Finally, we have the issue of different types of memory. Many excellent animal models of memory exist and are used routinely including contextual fear conditioning, water maze spatial learning, passive avoidance and odour-reward association. These tasks all have quite distinct sub-regional dependency within corticohippocampal circuitry and each have their own pros and cons. While much has been gleaned from the individual study of such paradigms, it is through the establishment of a universal role of a transcription factor or gene cluster across many or all such models as has been achieved for CREB that powerful insight into memory-associated synaptic plasticity is revealed.

Synaptic Plasticity-Associated Gene Clusters Identified by Microarray Approaches

The above caveats not with standing, several powerful microarray studies across a range of learning paradigms have begun to characterize the memory-associated synaptic plasticity transcriptome with key, core gene clusters emerging (Fig. **1**) [30,33,35,44]. Despite obvious differences in experimental design, including post-training time of analysis and learning task employed, some genes have consistently appeared in these studies. Several such genes, for example, microtubule-associated protein 1B and p38 mitogen activated protein kinase, were already know to be involved in memory from previous studies illustrating the validity of the microarray approach to identify novel memory-associated genes [45,46].

One of the first reports of the use of microarray approaches to examine memory-related transcripts came from Cavallaro and colleagues [30], who looked for gene expression changes in the HVI cerebellar lobule and hippocampus of the rabbit following training in the classical nictitating membrane response conditioning task. In this paradigm an innocuous stimulus, usually an auditory tone becomes associated with a noxious stimulus such as an air puff to the eye through repeated pairing. The conditioning task is fundamentally cerebellar dependent but also requires hippocampus to form the association if a time delay is introduced between the tone and air puff [47]. Cavallaro *et al.* [30] reported 79 and 17 gene expression changes 24h following training in the cerebellum and hippocampus, respectively, with the majority of genes exhibiting decreased expression. This early study was followed by a number of papers that shifted attention squarely to the hippocampus and looked at the classic hippocampal-dependant memory models, contextual fear conditioning, water maze and passive avoidance conditioning [48]. A classic spatial learning task, the Morris water maze was studied at 3 different time-points, 1, 6 and 24h, following the training session [31]. Over the three time points 140 genes were found to be differentially expressed relative to swim-time matched controls. Memory-related genes fell into defined categories such as cell signalling, including various growth factors, cell surface receptors and kinases; cell-cell interactions and cytoskeletal proteins, including microtubule associated proteins and adhesion molecules; and synaptic proteins.

In 2004 David Sweatt and colleagues conducted a microarray analysis of genes regulating at 4 time points up to 6h following contextual fear conditioning [33]. This group was the first to study discreet hippocampal subregions, investigating the CA1 and dentate gyrus separately. They observed co-regulating temporal waves of transcription in the two structures. Interestingly, they also reported physical grouping of regulating genes on chromosomes 10, 12, 16, 18, and 19, suggesting that the training event produces local epigenetic regulation to facilitate gene transcription [45]. Genes found within these chromosomal regions include transcription factors, growth factors, regulators of the actin cytoskeleton and signaling molecules [33].

Passive avoidance conditioning is another association task that is dependent on the ventral hippocampus [50]. In this task the animal learns an association between a dark chamber and a foot shock. This memory is robust, long-lasting and has a well defined time zero from which to time the events of consolidation. D'Agata and Cavallaro [44] used this paradigm to ask what gene expression changes occurred in the hippocampus 6h following the learning event. The study identified 38 such genes, again falling into functional classifications such as neurotransmission, growth/immune signalling, ion channels and second messenger cascades, mitogen-activated protein kinase (MAPK) in particular. This study also noted regulation of cytoskeletal and synaptic components such as microtubule associated protein (MAP) and synaptogyrin, the latter a key regulator of exocytosis and neurotransmitter release [51]. Again, there was a relatively even split between up- and down-regulated transcripts. O'Sullivan *et al.* [35] carried out a more extensive microarray analysis of gene expression following passive avoidance learning looking at multiple increasing time points up to and including 24h post-training. Also, rather than using whole hippocampal homogenate, the dentate gyrus subregion was studied as it represents the initial entry point for cortical information and it is a structure in which physical synaptic remodelling has been observed and temporally defined following passive avoidance learning [16]. This study showed that 597 genes were differentially expressed across the 6 time points studied relative to training naïve controls.

Taken together, these array data sets define several gene clusters at various points of the temporal cascade of consolidation. The known biology of members of such gene clusters suggest how they may contribute to the underlying mechanisms required for robust information storage (Figs. **1** and **2**).

Control of Transcription and Translation

Transcription inhibitor studies revealed two critical time windows for up-regulated gene expression, the first immediately following the learning event (0h post-training), the second during the early stage of memory consolidation, 2-6h post-training [22,23]. The first transcriptional wave, immediately following training, is now widely accepted to relate to increased expression of transcription factors and immediate early genes (IEGs) [23]. In our own studies, a quarter of all increased genes at the early times (0-2h) following passive avoidance learning fall into the transcription/translation functional group, with several coding specifically for transcription factors and IEGs [35]. One such gene, the CNS-specific transcription factor POU3f2, is known to regulate the intermediate filament protein nestin in cells undergoing neurogenesis, a process that is thought to play a role in some forms of memory formation [52-54]. The transcript for the ubiquitous translation factor Eif2b1 is up-regulated with POU3f2 perhaps in preparation for increased protein processing. Moreover, increased protein translation may well be coupled with enhanced mRNA processing as genes such as hnRNP show increased transcription in this early post-learning timeframe [35]. hnRNP is of particular interest as it is involved in the sequestration of mRNAs to neuronal dendrites facilitating local production of proteins in response to synaptic activity, a process much studied for its capacity to serve as a synapse tagging mechanism allowing selective, activity-dependent synaptic plasticity [55-57].

Selective Synapse Destabilization

Another component gene group regulating early in memory consolidation is cell adhesion molecules. Genes encoding cell adhesion regulators, such as CD9 and Thy1, are primarily down-regulated 2-6h post-training, representing ~20% of all down-regulated genes across this period of memory consolidation for passive avoidance [35]. Indeed, several mediators of de-adhesion, for example, G-protein alpha 12 (Gna12), shown to decrease E-cadherin-mediated cell-cell adhesion [58], are up-regulated at these early times. Together, these transcriptional events point to a period of synaptic destabilization 2-6h into memory consolidation. This idea is wholly consistent with previous observations that two adhesion molecules critical to synaptic integrity, APP and NCAM, are down-regulated at the protein expression level, likely through increased internalization and subsequent degradation, in this same early time period following passive avoidance training [11,12]. Moreover, perisynaptic regulation of protease activity is now well established to mediate the local degradation of cell adhesion molecules and extracellular matrix required for memory consolidation [13].

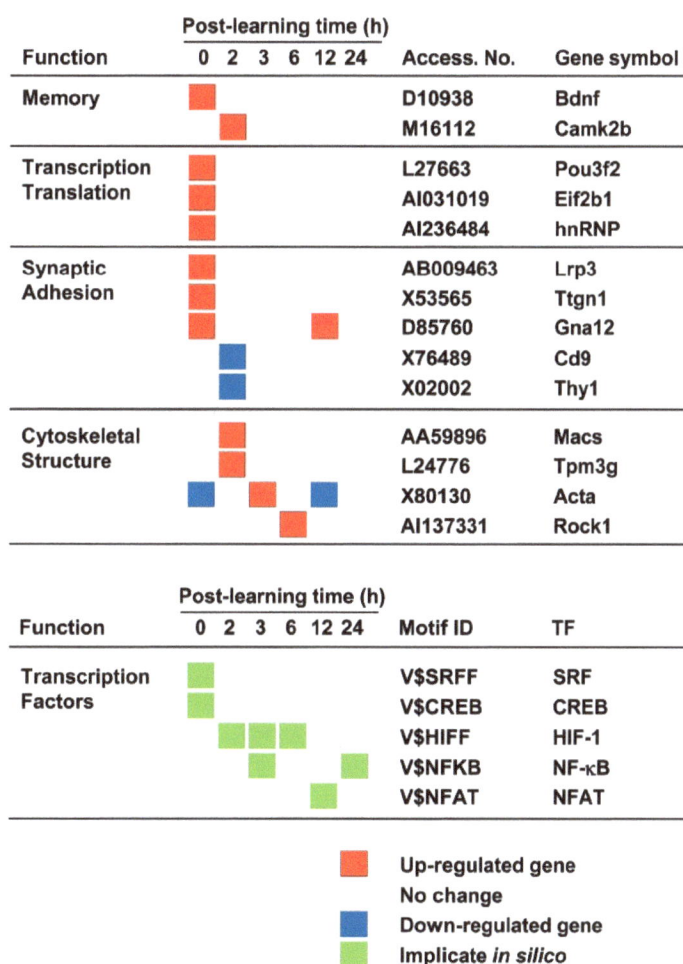

Post-learning time (h)

Function	0	2	3	6	12	24	Access. No.	Gene symbol
Memory	up						D10938	Bdnf
		up					M16112	Camk2b
Transcription Translation	up						L27663	Pou3f2
	up						AI031019	Eif2b1
	up						AI236484	hnRNP
Synaptic Adhesion	up						AB009463	Lrp3
	up						X53565	Ttgn1
	up			up			D85760	Gna12
		down					X76489	Cd9
		down					X02002	Thy1
Cytoskeletal Structure		up					AA59896	Macs
		up					L24776	Tpm3g
	down		up		down		X80130	Acta
				up			AI137331	Rock1

Post-learning time (h)

Function	0	2	3	6	12	24	Motif ID	TF
Transcription Factors	in silico						V$SRFF	SRF
	in silico	in silico					V$CREB	CREB
		in silico	in silico	in silico			V$HIFF	HIF-1
			in silico			in silico	V$NFKB	NF-κB
				in silico			V$NFAT	NFAT

Legend:
- 🟥 Up-regulated gene
- No change
- 🟦 Down-regulated gene
- 🟩 Implicate *in silico*

Fig. (2). Representative gene expression regulation in the hippocampal dentate gyrus in the 24h period following passive avoidance conditioning. This figure details the specific timing and nature of gene expression regulation for example genes from the major functional groups discussed in the text. The gene regulation data are taken from the microarray study of passive avoidance conditioning by O'Sullivan *et al.* [35]. The figure also shows the temporal involvement of transcription factors implicated in gene expression regulation by *in silico* analysis of the promoter regions of regulated genes (see Fig. **3** and text for further information).

The idea that adhesion molecules are actively stripped from the synaptic membrane during the early period post-learning finds support in the transcriptional up-regulation of genes encoding proteins such as low-density lipoprotein receptor-related protein 3 (Lrp3) and trans-Golgi network protein 1 (Ttgn1) [35]. The up-regulation of LRP mRNA has been shown to translate to the protein level and serve to increase the endocytosis of bulky trans-membrane proteins such as APP during this period following avoidance conditioning [12,59,60]. Moreover, inhibition of LRP-mediated internalization by the pan LRP inhibitor RAP renders animals amnesic but only when administered at the critical 2h post-training time [12]. The protein product of Ttgn1 functions in the shuttling of secreted and trans-membrane proteins between the trans-Golgi network, membrane and endosomes [63]. In particular, Ttgn1 is implicated in modulation of cell-extracellular matrix interaction at the synapse *via* rapid changes in the protein composition of the post-synaptic membrane [62-63]. Thus, gene expression control may contribute to the process of synaptic loosening which is presumably a necessary precursor to the elaboration of new synaptic connections during the subsequent 4-6hr post-learning time period of memory consolidation [11,16-18,64].

Cytoskeletal Structural Control

Gene expression regulation 2-6h post-training also putatively relates to the control of structural genes likely underpinning the long-lasting synaptic remodelling required for long-term memory formation [15,23-25,64]. In our studies, almost a quarter of genes up-regulated between 2-6h have a role in cellular structuring especially control of actin-cytoskeleton dynamics [35]. Moreover, actin is a prevalent cytoskeletal component of dendritic spines, precisely the structures observed to remodel during memory consolidation [16,17]. Blockade of actin polymerization has been shown to impair both LTP and contextual fear conditioning [61,62]. While real-time PCR could validate the transient up-regulation of alpha-actin mRNA at 3h following passive avoidance training, the levels of protein were not altered highlighting the importance of investigating microarray mRNA changes at the protein expression level. In the case of alpha actin it has been argued that lack of change at the protein level may not be surprising, however, as memory-associated actin remodeling would require both actin monomer addition and depolymerisation and degradation of existing actin structures. Moreover, two distinct pools of actin filaments, one stable the other unstable, have been identified in dendritic spines [67], the former being the cytoskeletal structure while the latter is the source of building blocks to support adaptive changes in synaptic connections [68]. Thus, during this critical period of memory consolidation, increased actin mRNA may facilitate generation of new actin protein to feed the unstable pool and to replace and remodel the degraded pre-existing spine cytoskeleton.

In terms of the signals controlling the cytoskeletal reorganization again evidence emerges from the transcriptional profiling approach. By way of example, myristoylated alanine-rich protein kinase C substrate (Macs), an actin-filament cross-linking protein regulated by PKC and calcium/calmodulin [69], is up-regulated at 2h post-learning. Developmentally, Macs is enriched in nerve growth cones where it regulates neurite outgrowth and synapse maturation. In adult brain, Macs expression is retained in regions associated with a high degree of plasticity while reducing Macs expression impairs spatial learning [70,71]. Also of interest in control of the actin cytoskeleton is the increase in Rho-associated coiled-coil containing protein kinase 1 (Rock1) mRNA at the 6h post-training time-point [35]. Rock can stabilize F-actin *via* two distinct mechanisms, increased phosphorylation of LIM kinase (LimK) or enhanced actin-myosin interaction achieved through inhibition of myosin light chain phosphorylation [68.69]. Moreover, a direct role of Rock in memory formation has been demonstrated as intrahippocampal infusion of a Rock inhibitor impairs long-term memory [74]. The timing of the up-regulation in Rock1 is again telling as this period corresponds with the appearance of new synaptic connections requiring stabilization of newly generated actin cytoskeletal structures [16-18].

Transcription Factors in Memory-Associated Synaptic Plasticity

In silico Transcription Factor Binding Site (TFBS) Analysis

The preceding section shows that gene regulations identified through microarray approaches have the clear potential to flag molecular players required for long-term memory formation. Moreover, inherent in these studies but to date relatively untapped is the identification of co-regulating gene clusters, the existence of which implicate gene expression regulation through altered activity of specific transcription factor modules. Powerful analysis techniques are now emerging that allow the promoter structures across members of such gene clusters to be probed, identifying likely controlling transcription factors (Figs. **2** and **3**).

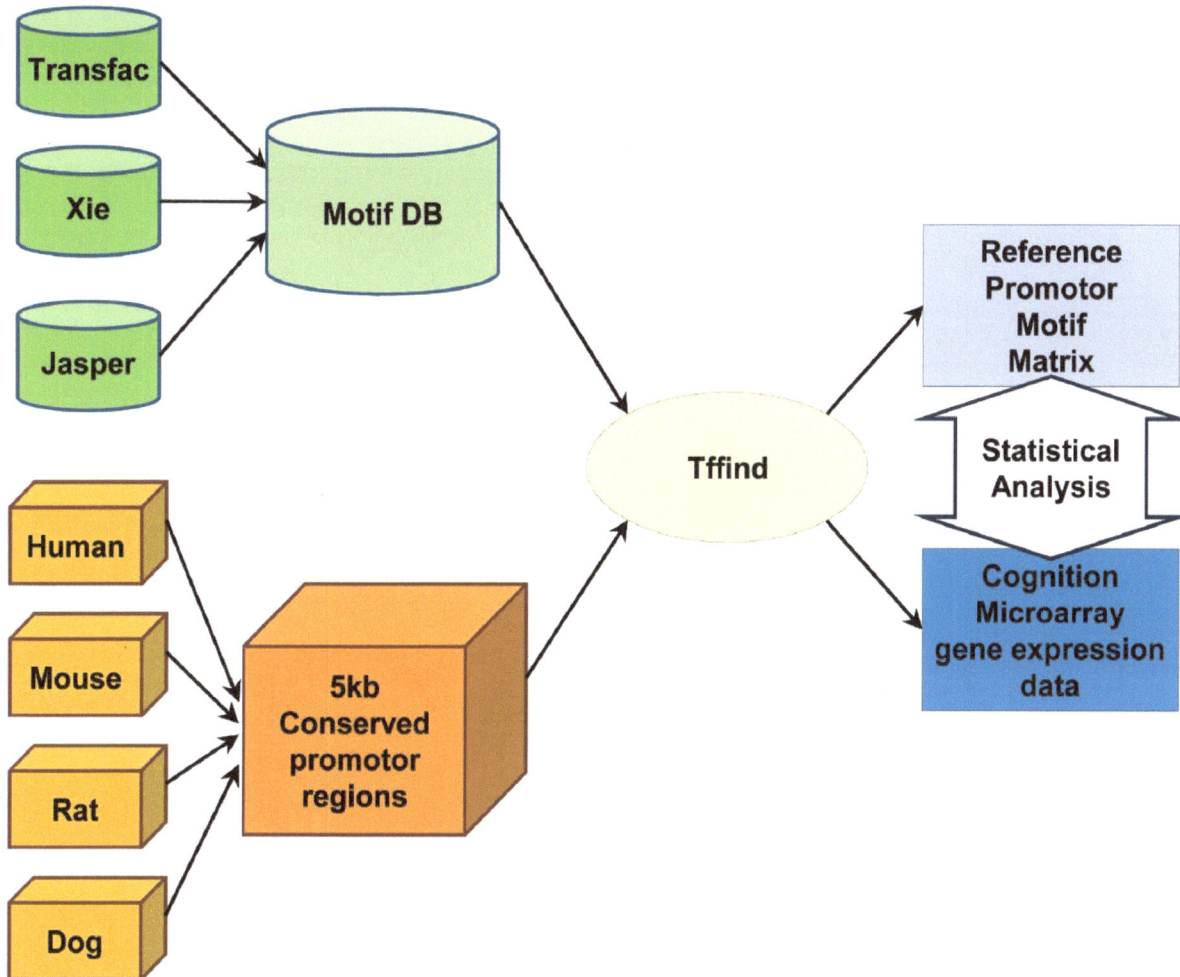

Fig. (3). Methodology used to identify over-represented transcription factor binding sites. The tffind component of the Multipipmaker package is used to identify the regions conserved between human, mouse, rat and dog in gene promoters within 4kb of the transcriptional start site. Using the combined motif library from Transfac, Jaspar and Xie, conserved promoter regions can be scanned to identify transcription factor binding motifs present. This generates a matrix of the number of motifs present in the promoter of each gene for 1217 motifs and 13551 probesets. Subsequently, one can quickly determine which of the motifs are over-represented in a list of differentially expressed genes at any time-point post-learning by the Fisher exact test.

Transcription is controlled by transcription factors, proteins which support or inhibit the formation of a transcriptional pre-initiation complex at a target gene [21,75]. Most commonly, transcription factors act by recognising and binding to short (5-15 nucleotide bases), *cis*-active motifs in the promoter region of genes and thus regulate the expression of a host of specific target sequences [76,77]. Logically, groups of genes observed to co-regulate temporally are likely to share regulatory control. Thus, co-regulating genes following learning can be studied to learn more about the transcriptional control at play at various times in the memory-associated synaptic plasticity process. Two of the microarray studies discussed above used consolidation-associated genes together with bioinformatic analysis to identify over-represented transcription factor binding sites (TFBSs) [33,35]. The transcription factors c-Rel and NF-κB implicated by these studies were proven, in independent experiments, to have a role in memory consolidation.

We and others have developed methodology for such *in silico* analysis of gene promoters to identify TFBSs over-represented in a gene cluster (Fig. **3**); indeed several powerful software platforms to this end are now commercially available. Our latest iteration of this approach employs the tffind component of the Multipipmaker package [78] to identify the regions conserved between human, mouse, rat and dog in gene promoters within 4kb of the transcriptional start site [79]. Using the combined motif library from Transfac, Jaspar and Xie, conserved promoter regions can be scanned to identify transcription factor binding motifs present [79-81]. This generates a matrix of the number of motifs present in the promoter of each gene for 1217 motifs and 13551 probesets. Subsequently, one can quickly determine which of the motifs are over-represented in a list of differentially expressed genes at any time-point post-learning by the Fisher exact test.

Analyses of memory-associated synaptic plasticity transcriptomes by the bioinformatics approach discussed above implicates NF-κB and CREB in gene expression regulation following learning, offering good support that other TFBSs identified through this approach represent solid candidates for the title 'memory-associated transcription factor' [33,35]. The timing and identity of some of the implicated transcription factors begin to create a compelling transcriptional cascade across the key phases of memory-associated synaptic plasticity (Fig. **2**). CREB and serum response factor (SRF) are over-represented in the immediate post-training period implying an involvement in transcription/translation and synapse destabilization. NF-κB is implicated 2-3h post-learning when new synaptic structures are emerging. HIF-1, well known for its role in hypoxia, is also implicated in gene regulation during this middle phase of synaptic reorganization. In the later phase of selective synapse stabilization we find a putative role for nuclear factor of activated T-cells (NFAT) and, later again, c-Rel. As this late phase is marked by the two competing processes of synapse stabilization of connections to be retained and dismantling of synapses no longer required, these transcription factors might contribute to either or both of these crucial components of memory-associated synaptic plasticity. Below we discuss the known biology of these transcription factors offering rationale for the nature of their specific roles in the context of memory-associated synaptic plasticity.

Serum Response Factor - SRF

SRF is a widely distributed, evolutionarily conserved factor first identified by its ability to drive serum-activated expression of IEGs such as c-fos in replicating cells [82]. SRF is a member of the MADS-box transcription factor family, which binds as a homodimer to the CArG-box (CC[A/T]6GG) consensus sequence, referred to as the serum response element (SRE) [83]. While binding to the CArG-box is sufficient to mediate SRF-dependent transcription, the promoter regions of many genes contain one or more SRE, typically comprising a CArG-box and an adjacent *cis* element [84]. These sites act as assembly locations for multiprotein complexes with SRF dimers in response to activation by at least two alternative pathways: the mitogenic pathway involving proteins of the ternary complex factor (TCF) family, and the Rho-GTPase pathway involving myocardin-related transcription factor (MRTF) proteins. SRF is also reported to functionally interact with other transcription factors including Sp1, ATF6, GATA4 and Nkx2.5, mainly in the context of cardiac cells [85-87]. Thus, while SRF can control expression of many genes and regulate many different biological functions, the specific repertoire of genes induced by a particular stimulus is largely dictated by the interaction of SRF with either the TCF or MRTF family cofactors.

The possibility that SRF is involved in learning and synaptic plasticity remains poorly studied. For a long time *in vivo* studies were hampered as SRF knockout (KO) mice were not viable, undergoing early embryonic lethality at embryonic day 6 [88]. The generation of conditional KOs has pointed to a clear role for SRF at various stages of neuronal development. SRF deletions in the developing and perinatal nervous system results in impaired outgrowth and connectivity patterning in the rostral migratory stream and mossy fibre pathway, respectively [89,90]. In the adult hippocampus, deletion of SRF results in mice with impaired activity-dependent induction of IEG expression and synaptic plasticity [91]. Interestingly, adult hippocampus lacking SRF expression has significant deficits in both early and late phases of LTP, indicating that SRF contributes to transcription events immediately following activity, likely *via* the IEGs, and also later transcriptional events. More recently, SRF decoys have been used to inhibit individual subunit transcription factor function [92]. Decoys are double-stranded DNA oligonucleotides that competitively bind specifically to the transcription factor of interest, blocking expression of target genes [93]. Intra-hippocampal administration of SRF decoy, but not a scrambled sequence control, has been shown to inhibit spatial memory [94]. These studies provide direct evidence for a role for the activity of SRF in the regulation of memory consolidation.

SRF and Immediate Early Genes

Interaction of SRF with the TCF family of proteins is well established as a mechanism of eliciting a transient and rapid regulation of IEG expression. Indeed, TCFs were first discovered, in HeLa cell nuclear extracts, located on the serum response element of the c-fos promoter [95]. The TCF family is known to have three members: Elk-1 (Ets-like protein 1), Sap-1 (SRF accessory protein 1) and Sap-2 (also known as Net/ ERP/ Elk-3) [95-98]. These TCFs share five highly conserved domains which mediate DNA binding (the N-terminal ETS domain), SRF interaction (B box), MEK interaction (D and F box) and transactivation (C box) [95,99,100].

In neurons, SRF-TCF binding was found to mediate the c-fos transcriptional response following glutamate receptor activation and, more recently, BDNF stimulation [101-103]. This signal is transduced *via* the MAPK/ extracellular signal-regulated kinase (ERK) kinase (MEK) pathway, initiated by receptor activity-driven calcium influx. Specifically, activated MEK phosphorylates two serine residues on the TCF Elk-1, Ser^{383} and Ser^{389}, causing it to undergo a conformational shift leading to increased DNA-binding and transcriptional activity. Importantly, while this activation requires MEK signaling and is blocked by the MEK specific inhibitor U0126, it does not require actin polymerization, another signaling pathway modulating SRF expression discussed further below [104]. The function of the SRF-TCF induced gene expression following CNS synaptic activity is primarily the up-regulation of a host of IEGs including c-fos, Egr1, Egr2 and SRF itself. These transcription factors in turn initiate a cascade of transcriptional activity of genes encoding growth factors (e.g. BDNF), cytoskeletal proteins (e.g. CAM1), synaptic proteins (e.g. PSD95) and proteins involved in neurotransmission (e.g. voltage gated channels) [105,106]. The protein products of these genes likely play a role in receptor internalization or clustering, signaling molecules and synaptic restructuring. In support of this, inhibition of SRF-TCF signaling by mutation of Elk-1 disrupts dendritic growth in primary striatal neurons [107].

SRF and Structural Regulation

The other major family of SRF coactivators are the MRTF family, which include myocardin, megakaryoblastic acute leukaemia 1 (Mkl1; also referred to as MAL1 and MRTF-A), and Mkl2 [108,109]. Of these, Mkl1 and Mkl2 are ubiquitously expressed with prominent expression in the hippocampus and other brain structures, while expression of myocardin is restricted to smooth and cardiac muscle [110,111]. Though they have somewhat distinct expression there is functional redundancy between Mkl1 and Mkl2 as complete blockade of SRE-dependent reporter gene activity requires inhibition of both Mkl proteins [112]. A conserved domain on Mkl proteins, the B-box, is essential for their interaction with SRF and the neighbouring Q-box facilitates the Mkl-SRF complex formation [108,109]. Interestingly, point mutation analysis has shown that both MRTF and TCF cofactors bind SRF at the same site of the SRF protein [113]. As such, in tissues in which both MRTFs and TCFs are expressed competition for SRF would be expected. This is borne out by biochemical studies showing that Elk-1 suppresses the activity of Mkl-SRF down-stream genes [114]. Therefore, as neurons express both MRTFs and TCFs, the relative nuclear abundance of these proteins likely controls the nature of the transcription factor complex assembled and thus identity of target genes expressed.

In contrast to SRF-TCF activity, SRF-MRTF activity is MEK-independent regulated instead through Rho-family GTPases including RhoA, Rac and Cdc42 [115]. This pathway depends on the actin treadmilling cycle, in which monomeric G–actin is added to and disassembled from the barbed and pointed ends of filamentous F-actin, respectively [116]. MRTF proteins possess three N-terminal RPEL domains, which mediate binding to G-actin and determine the subcellular localisation and regulation of Mkl1 and Mkl2 [117]. Increased actin polymerisation through activity of Rho family members induces translocation of Mkl proteins from the cytoplasm to the nucleus in NIH 3T3 cells and cortical neurons [109,118]. Once accumulated within the nucleus, Mkl proteins associate with SRF and stimulate SRF-dependent transcription of target genes, including α-smooth muscle actin (acta), caldesmon, and tropomyosin, as well as IEGs such as c-fos and SRF itself [119-121]. It should be noted however, that there exists some discrepancy as to the subcellular localisation of Mkl1 in neurons with some groups reporting largely cytoplasmic, and others largely nuclear, localisation [111,122,123]. It is unclear at this point whether this represents differences in cell populations or simply the antibodies used in these immunohistochemical studies.

Findings from many different groups have revealed that Rho-activated SRF-induced target genes comprise an impressive list of genes with functions related to actin dynamics. These include actin filament-associated genes, filament reorganization genes, and integrin coupling-associated genes [124]. Given that the actin cytoskeleton is both an upstream regulator of Mkl activity and a down-stream target of Mkl-SRF, it seems clear that there exists a feedback loop for SRF-dependent control of actin cytoskeleton-related processes. Looking strictly at MRTF proteins in the CNS, Mkl1 translocation has been reported following synaptic activation by NMDAR stimulation and KCl-induced membrane depolarisation in cultured cortical neurons, raising the possibility that Mkl proteins contribute to SRF-driven transcription underlying neuronal plasticity [122]. Supporting this, dominant negative Mkl1 mutants and actin point mutations, which cannot bind Mkl proteins, have impaired dendritic growth in primary neuronal cultures [90,125,126].

Recent findings using passive avoidance learning suggest that regulation of Mkl1/2 nuclear accumulation contributes to 2 waves of SRF-dependent gene regulation during memory consolidation [123]. The first wave immediately post-training relates to IEGs and is facilitated by Mkl down-regulation and likely release of TCF-SRF activity. A second transcriptional wave occurs later at the 3 h post-training time point includes known SRF-responsive structural genes tropomyosin 3 and alpha smooth muscle actin and is accompanied by Mkl nuclear accumulation in hippocampal neurons.

cAMP Response Element-Binding Protein - CREB

The best studied transcription factor in the context of hippocampal plasticity is the cAMP response element-binding protein (CREB), and it is reviewed extensively in other chapters of this publication. Briefly, CREB is a member of the bZIP-type family of transcription factors which also includes ATF1, Jun, Fos and C/EBP [127]. CREB dimers bind to conserved cAMP-responsive elements (CRE), a TFBS characterised by the TGACGTCA consensus sequence. Activated CREB dimers, phosphoylated at serine 133, bind to CRE promoters and initiate the formation of a larger transcriptional complex by recruiting proteins such as CREB-binding protein (CBP). CBP can then activate transcription *via* its intrinsic histone acetyltransferase activity, which acts to 'open' nearby chromatin and allow RNA polymerases to initiate transcription [128]. Several signaling pathways are believed to regulate CREB activity in response to receptor-mediated actions within hippocampal neurons. For example, increased levels of Ca^{2+} and cAMP activate Ca^{2+}/calmodulin-dependent kinase IV (CaMKIV) and cAMP protein kinase (PKA), respectively, stimulating CREB-driven transcription [129,130]. The MAPK pathway has been extensively implicated in long-term memory with MAPK inhibitors blocking CREB phosphoylation and impairing hippocampal learning [131-133]. More recent studies have shown training-induced activation of MAPK, down-stream CREB kinases and CREB within individual hippocampal neurons [134]. Interestingly, mutations in the human MAPK-activated CREB kinase gene RSK2 lead to a form of mental retardation, Coffin-Lowry Syndrome, further evidencing the crucial role of this pathway in normal brain function [135].

CREB is a crucial transcription factor in the regulation of neuroplasticity and its activity is required for long-term facilitation in *Aplysia*, long-lasting LTP in hippocampal slices, and long-term memory in *Drosophila* and rodents [20,25,136-138]. LacZ reporter studies and direct measurement of Ser[133] phosphorylated CREB show increased CREB activity following training in hippocampal-dependent tasks [139,140]. In addition, suppression of CREB signaling by antisense oligonucleotides to CREB impairs consolidation, but not acquisition, of spatial learning [141,142]. CREB activity is reported to regulate the expression of over 100 genes, a list that includes transcription factors (e.g. c-fos), growth factors (e.g. BDNF), and neurotransmitter receptor subunits (e.g. Gria1) [143,144]. However, it should be noted that CREB activity at target genes can be complex and does not act merely as an 'ON' switch. BDNF expression, for example, can be selectively regulated in distinct neuronal populations and chromatin immuneprecipitation assays confirmed that the identity of CREB target genes differs from one cell type to another [145,146]. Recently, CREB activity has been shown to regulate the expression of a microRNA, miR132 [147,148]. miRNAs encode short double-stranded RNAs, which mediate gene silencing by binding to target mRNA and mediating translational repression or mRNA degradation [149]. miR132 targets the Rho family GTPase-activating protein p250GAP, inhibiting its translation, and thereby limits activity-stimulated dendritic growth. This is a novel mechanism of CREB-mediated control of neuroplasticity and illustrates that there remains much to be discovered about this extensively studied transcription factor.

Nuclear Factor-κB - NF-κB

Nuclear factor-κB (NF-κB) was originally identified as a key transcription factor involved in the regulation of cytokine production [150]. NF-κB consists of two subunits selected from a family of five (p50, p52, p65, c-Rel and RelB), with the p50/p65 heterodimer by far the most abundant form. In neurons, NF-κB exists in two pools, a latent pool, sequestered inactive in the cytoplasm bound to its inhibitory chaperone IκB, and an active pool in the nucleus where it functions to regulate gene expression. In order for latent NF-κB to become activated, IκB must first be phosphorylated, which causes its dissociation from NF-κB and targets IκB for degradation by the ubiquitin 26S proteosome system [151]. Free from the inhibitory constraints of IκB, NF-κB translocates to the nucleus, where it binds to specific TFBSs and, together with adjacent enhancer elements, modulates the expression of a whole host of genes that include IκB thus providing an auto-regulatory loop [152,153]. NF-κB has long been known to be widely expressed in the CNS but it has only recently been implicated in memory processing [154].

There is now clear evidence that several of the NF-κB subunits are required for various types of hippocampus-dependant cognitive function. Early work by Meberg *et al.* [155] found up-regulated levels of NF-κB 60 min post-high frequency stimulation (HFS). These findings were later supported by Freudenthal *et al.* [156], who demonstrated that the p65 subunit of NF-κB becomes activated throughout the hippocampus *in vivo* 15 min post-high frequency stimulation. Interestingly, both studies demonstrated that LTD-inducing low-frequency stimulation (LFS) also up-regulated NF-κB levels but to a lesser extent than strong high-frequency synaptic activation supporting the concept that NF-κB oscillations within activated neurons may function to translate the extent of recent synaptic activation into a meaningful downstream signal with this activity in turn serving as a memory of the metaplastic state of potentiated/depressed synapses.

NF-κB has molecular properties and cellular localisation that place it in an ideal position to convert recent synaptic activity into a meaningful transcriptional program driving subsequent long-term plastic changes associated with memory formation [155,157-160]. For example, Kaltschmidt and colleagues used eGFP-labelled p65 to monitor NF-κB cellular redistribution in hippocampal neurons following activating stimuli showing that depolarisation drives translocation of NF-κB from neurites to the nucleus *via* dynein-mediated transport along microtubules [159,161]. Moreover, disruption of microtubule function by colchicine or vincristine, drugs which induce cognitive deficits when infused into the hippocampus, reduced NF-κB-dependent transcription activity [161-163].

Mice expressing the so-called IκBα super-repressor, a nondegradable inhibitor of NF-κB/Rel activation, in forebrain neurons show impairment in both acquisition and consolidation of the Morris water maze [164]. More recently, intra-hippocampal administration of NF-κB decoys inhibited both spatial and avoidance memory [29,94]. As mentioned above, bioinformatics analyses of microarray data on learning-induced gene expression implicated NF-κB subunits c-Rel and p65 activation during memory consolidation [33,35]. In support of this suggestion, individual knock-outs of p65 and c-Rel subunits are known to produce significant deficits in radial arm maze and contextual fear conditioning, respectively [33,160]. Taken together, these studies suggest that NF-κB acts to transcriptionally regulate gene expression during memory consolidation across a wide range of learning paradigms.

In the CNS, NF-κB is known to be activated by a variety of stimuli involved in neuronal plasticity. The best characterised of these is the activation of NF-κB by the neurotrophin nerve growth factor (NGF). NF-κB subunits have also been shown to activate in response to synaptic transmission, binding of glutamate to its cognate receptors, and increased calcium *in vitro,* as well as following hippocampal learning *in vivo* [35,160,165]. In addition, while our current knowledge of the genes regulated by NF-κB comes mostly from studies of non-neuronal cell types, a number of genes with direct relevance for the nervous system have been described and begin to elucidate the possible function in the context of neuronal plasticity. NF-κB regulates the expression of several cell adhesion molecules and these may well contribute to its effect on the size and complexity of dendritic arbours. NCAM, one such NF-κB target gene, mediates dynamic cell-cell interactions during activity-dependent synaptic plasticity and blockade of its normal activity disrupts hippocampal learning [11,164,165]. Sequence analysis of the NCAM promoter initially revealed a potential NF-κB binding site [168] and NF-κB has been shown to bind to the promoter region and increase expression levels of NCAM in response to nitric oxide [169]. Other pertinent genes containing NF-κB binding sites in their promoter regions include cell adhesion molecules such as APP, Tenascin-C, and β1 integrin [170-172], as well as several glutamate receptors, Grin2A, Grin1 and mGluR2 (a full list of NF-κB target genes is updated at: http://people.bu.edu/gilmore/nf-kb/index.html) [173-175]. NF-κB binding sites are present in one third (16) of all the genes regulating at 3h following passive avoidance learning with half of these genes functioning in cell adhesion and structure control [176]. Specifically, Slit and Trk-like family member 1 (Slitrk1) and T-lymphoma and metastasis 1 (Tiam1), both expressed in the hippocampus and known to play a role in neuronal growth [177-179] were shown to possess NF-κB binding sites in their promoters and, moreover, were significantly up-regulated 3h following learning [176] suggesting a role for NF-κB in memory-associated structural plasticity.

Recently, immunofluorescent labelling of the active NF-κB p65 subunit has allowed the direct detection of the activation and translocation of NF-κB. This technique has allowed our group to identify increased nuclear localization of NF-κB in dentate granular neurons and CA pyramidal neurons in both the dorsal and ventral hippocampus 3h following passive avoidance training [35,176]. Basal expression of NF-κB differs markedly across the different neuronal subpopulations of the adult rat hippocampus [180]. CA1 and dentate granule neurons express similar levels of constitutively active NF-κB in both their nuclear and cytoplasmic compartments. CA3 pyramidal cells, however, exhibit higher basal concentrations of activated p65. Interestingly, immature neurons located in the neurogenic zone of the dentate gyrus exhibit relatively low levels of activated NF-κB. In this regard, it is important to recognise that cell-cell contacts mediated by NCAM may contribute to greater NF-κB activity in the nuclei of mature hippocampal neuronal populations. NCAM-mediated adhesion, a feature of mature synaptic connections, up-regulates NF-κB activity through kinase signalling cascades similar to those employed by cytokines [181]. The immaturity of cells located within the infragranular layers of the dentate in conjunction with high levels of polysialic acid (PSA) expressed on the surface of these cells, a post-translational modification of NCAM that destabilises NCAM-mediated adhesion, may contribute to the low basal nuclear NF-κB activity in this highly neuroplastic subpopulation of dentate cells. Thus, NF-κB may contribute specifically to synaptic plasticity in mature neurons.

A Novel Role for HIF-1 in Memory-Associated Synaptic Plasticity

As discussed above, a significant limiting factor in our steps toward elucidating the molecular and cellular mechanisms underlying memory formation and consolidation is the clear need to focus on the relatively small fraction of cells that are involved in laying down the memory trace. Techniques such as electrophoretic mobility shift assays (EMSA), western blotting, ELISA and 2D-gel electrophoresis routinely employed to measure changes in specific protein levels following learning [182-184] involve the indiscriminate analysis of all cells in the sample, a significant dilution of any memory-specific protein-level change. Approaches to purify protein samples to contain more specifically cell populations of interest include separating dentate and CA structures in the hippocampus or using laser microdissection techniques to cut out the neuronal cell lines of interest [185]. A key method that circumvents this dilution problem, however, is immunohistochemical imaging that allows vastly superior spatial resolution in differentiating the responses of distinct cell types to a specific memory-related task. For example, one can measure the changes in a specific protein in neuronal versus glial cell types in the different regions of the hippocampus following a learning event. Using this technique, we present data here supporting a potentially novel role for the transcription factor, hypoxia-inducible factor-1α (HIF-1α) in hippocampal-dependent spatial learning.

HIF-1α is a ubiquitously expressed transcription factor predominantly studied in the context of decreased tissue oxygen levels or hypoxia [186]. HIF-1α is thought to remain inactive in the cytoplasm in normoxic environmental conditions through the action of prolyl hydroxylases (PHDs) which hydroxylate two conserved proline residues on HIF-1α. Under hypoxic conditions, however, PHDs are inactivated as a result of the decrease in oxygen availability in the cell allowing translocation of HIF-1α to the nucleus where it binds in a heterodimeric fashion to HIF-1β/Arnt on hypoxia response elements (HREs) and regulates the transcription of a number of genes including vascular endothelial growth factor (VEGF), erythropoietin (Epo) and transforming growth factor beta (TGFβ) [187]. HIF-1α has nevertheless been shown to be activated under normoxic conditions, most notably by cytokines such as TNFα [188].

Our microarray data from O'Sullivan *et al.* [35] and an as yet unpublished study of water maze learning indicated that several genes exhibiting learning-specific alterations in their dentate expression levels at early time-points following hippocampal-dependent forms of learning have an over-representation of HIF-1 binding sites in their promoter regions. We therefore quantified changes in HIF-1α protein levels in the CA1, CA3 and dentate gyrus of rats trained in the water maze task using immunofluorescence and image analysis techniques. Immunofluorescent staining for HIF-1α in the adult rat hippocampus revealed a high level of expression in endothelial and glial cells.

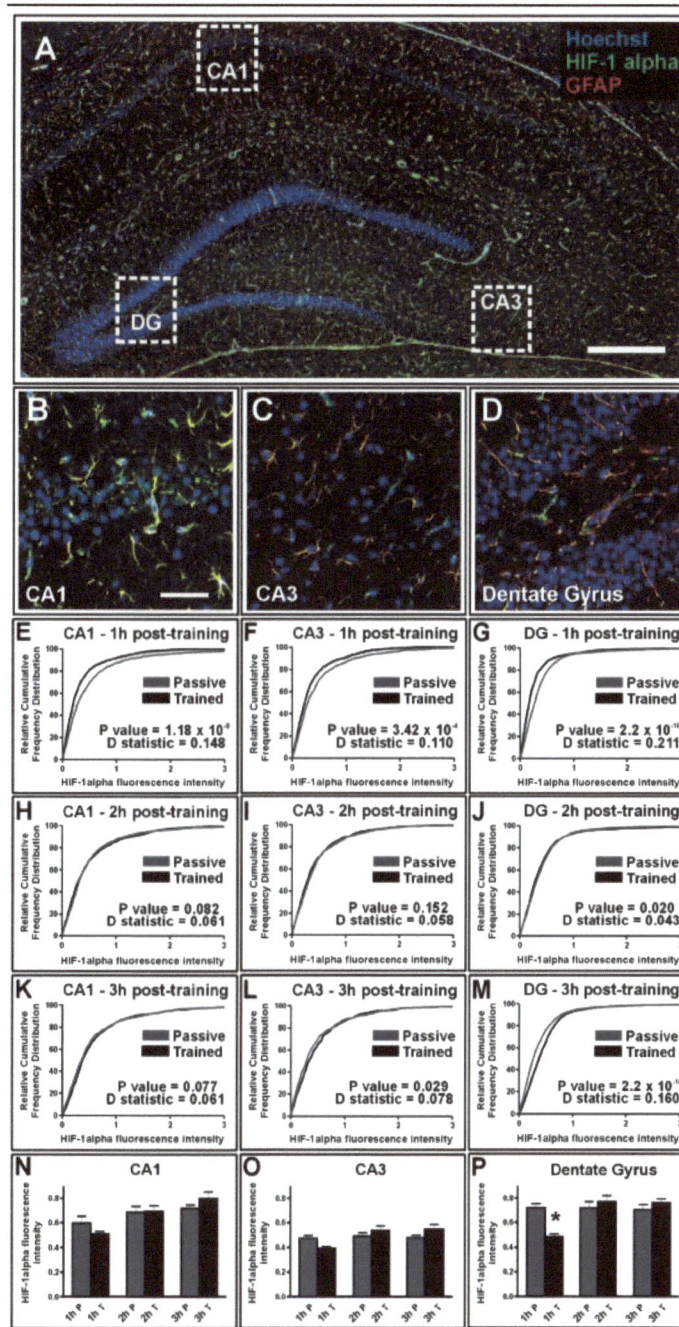

Fig. (4). Change in HIF-1α protein expression following water maze training in the rat dorsal hippocampus. A: Immunohistochemical montage showing HIF-1α expression in the dorsal hippocampus (–3.3 mm from Bregma) of the adult (post-natal day 80) rat. HIF-1α is labelled green (FITC-conjugated secondary antibody). Red staining represents glial-fibrillary acidic protein (GFAP)-labelled astrocytic processes. Nuclei are counterstained blue using Hoechst 33258. Boxes indicate the neuronal cell populations analysed. Scale bar = 250 μm. B – D: High-magnification images of the CA1, CA3 and dentate gyrus (DG) regions, respectively. Scale bar = 50 μm. E – G: Changes in nuclear HIF-1α activity 1h post-water maze training in the CA1, CA3 and DG in trained versus passive control animals. Graphs represent cumulative frequency distributions of each population of cells. Differences in the distributions were analysed using the Kolmogorov-Smirnov (K-S) test. A p value of < 0.01 was deemed significant and the difference (D) statistic is a measure of how divergent passive and trained distributions are from one another. H – I: Changes in nuclear HIF-1α activity 2h post-water maze training in the CA1, CA3 and DG in trained versus passive control animals. K – M: Changes in nuclear HIF-1α activity 3h post-water maze training in the CA1, CA3 and DG in trained versus passive control animals. N – P: Changes in HIF-1α activity in GFAP-labelled astrocytic processes 1, 2 and 3h post-water maze training in the CA1, CA3 and DG in trained versus passive control animals. Results were analysed using one-way ANOVA and Bonferroni post-hoc tests. A p value of < 0.01 was deemed significant.

Subsequent double-staining for HIF-1α and GFAP (Fig. **4**), the classic astrocytic marker, confirmed that the majority of GFAP-positive processes expressed HIF-1α. We quantified both GFAP-associated HIF-1α levels and nuclear HIF-1α activity in the CA1 and CA3 pyramidal layer and dentate granule neurons of the dorsal hippocampus at three time-points early in the memory consolidation process, i.e. 1, 2 and 3h. All three regions exhibited a significant, training-specific decrease in nuclear HIF-1α expression 1h post-learning compared to swim time-matched control animals (Fig. **4**). This down-regulation was also evident in glial processes but only in the dentate gyrus (Fig. **4**). While nuclear and astrocytic HIF-1α activity had returned to control levels at 2h post-training, there were up-regulations in nuclear HIF-1α specifically localised to the dentate gyrus one hour later, at the 3h post-spatial learning time-point.

These learning-specific alterations in HIF-1α nuclear expression point to a role for HIF-1α-regulated gene products in spatial memory formation. As can be revealed by analyses of the cumulative frequency distributions, learning-specific changes in HIF-1α expression occurred in a large proportion of the cells (Fig. **4**). Moreover, conventional methods that calculate the mean fluorescence of cell populations mostly failed to identify HIF-1α regulations likely due to large variation in expression across the cell population *per se*. The changes in HIF-1α activity might occur as a result of regional alterations in oxygen availability, reflective of modulations in hippocampus neuronal activity and metabolic demand. The temporal pattern of HIF-1α activity 1 and 3h post-water maze training may reveal periods of high and low oxygen supply, respectively, during the consolidation process for spatial memory. Alternatively, in the context of memory, cytokines such as TNFα might be regulating HIF-1α signalling independent of oxygen tone. The question remains as to whether the resultant gene regulation is functional in the context of the memory trace. Certainly, VEGF and TGFβ, both HIF-1α targets that regulate at the message level following learning, initiate signaling cascades capable of participating in the synaptic reorganization required for effective memory consolidation.

Nuclear Factor of Activated T-Cells - NFAT

Another family of transcription factors initially studied for their role in immune responses but now of interest in the context of memory-associated neuronal reorganisation is the nuclear factor of activated T-cells (NFAT) group. This family currently comprises five proteins containing the rel DNA binding domain, four of which (NFAT1-4) contain the calcium-sensing nuclear localization signal (NLS) domain [189,190]. These Ca^{2+}–regulated isoforms are activated by increases in intracellular calcium *via* the phosphatase calcineurin. Calcineurin (CaN) partially dephosphorylates NFAT, exposing its NLS, and promoting translocation of the transcription factor from the cytoplasm to the nucleus [189,191,192]. Once in the nucleus, NFAT binds to and initiates transcription with one of a number of nuclear partners, generically termed NFATn [193,194].

Several lines of evidence support a role for NFAT signalling in activity-dependent plasticity. Firstly, NFAT undergoes nuclear localisation and initiates transcription in response to several synaptic stimuli [194]. In these studies, luciferase-based gene reporter assays showed NFAT3-dependent transcription in hippocampal neurons to be activated by the neurotrophins BDNF and NGF, NMDA receptor stimulation and K^+-mediated depolarisation. Secondly, NFAT has been implicated by at least one high-throughput study as a regulator of gene expression following hippocampal learning [35]. Importantly, CaN, the regulator of NFAT, is required for NMDA-dependant LTD [195] while inhibition of CaN enhances hippocampus-dependent long-term memory formation [196,197]. These studies suggest that CaN acts to constrain processes involved in LTP-like synaptic plasticity and memory formation, and perhaps offer insight into the specific role of NFAT transcriptional activity during memory consolidation. Until recently, this question was difficult to answer as the target genes of NFAT in the CNS were largely unknown. The type 1 inositol-1,4,5-triphosphate receptor (IP3R1), which regulates the release of calcium from intracellular stores, was the only gene identified as being regulated by NFAT-dependent transcription within neurons [193,194]. This was proposed to act as a positive feedback loop to enhance synaptic calcium signalling [194]. However, NFAT3 has recently been shown to bind to the promoters and regulate the expression of the AMPA glutamate receptor subunit 2 (GluR2) and the axon outgrowth and guidance molecule GAP-43 in neurons, both known to play a central role in the control of synaptic strength during learning and memory [198,199]. Specifically, over-expression of NFAT3 represses GAP-43 reporter activity in primary cortical neurons, suggesting that NFAT3 is a negative regulator of this axonal outgrowth molecule [199].

Studies of developing neurons are proving very useful in helping to elucidate the function of this transcription factor and they have clearly revealed a role for NFATs in the control of axon growth. Triple NFAT1/3/4 mutants show severe defects in the axonal projections of sensory neurons and crucially neuronal explants from these triple mutants were used to demonstrate that NFAT-mediated axonal outgrowth is a result of neurotrophin and netrin signalling [200]. Unexpectedly however, recent elegant work in the developing neural circuit of *Xenopus* showed that blockade of CaN, thereby inhibiting the nuclear localisation of NFAT, increases dendritic branching and branch dynamics [201]. Together, these findings show that activation of NFAT-dependent transcription is a key regulator of dendritic and synaptic restructuring, specifically *via* dendritic retraction. Thus, given its suggested involvement at later times post-training [35], NFAT could function to terminate activity-induced synaptic growth and to co-ordinate the required synaptic elimination following a learning event.

CONCLUSIONS

Transcriptomics has already provided a wealth of new information about the molecular underpinnings of memory-associated synaptic plasticity. The identification of genes regulating at specific times in consolidation has allowed insight into the sequential processes at play during the key phases of synaptic destabilization, new synapse construction and selective synapse retention. In particular, these gene clusters have allowed the putative identification of memory-associated transcription factors that likely control transcription across the 24h post-learning period. Substantial future research initiatives will be needed around each of these putative transcription regulators to fully elucidate the nature of their role in memory-associated synaptic plasticity. The value of these endeavours cannot be overstated as a deep understanding of memory-associated synaptic plasticity will certainly guide the development of improved treatments of disease states of brain function such as Alzheimer's, schizophrenia and depression.

MATERIALS AND METHODS

Animal Maintenance and Behavioural Assessment in the Open Field Arena

Postnatal day 80 male Wistar rats (330–380 g) were obtained from the Biomedical Facility at University College Dublin, Ireland. All experimental procedures were approved by the Animal Research Ethics Committee of the Biomedical Facility at UCD and were carried out by individuals who held the appropriate licence issued by the Minister for Health and Children. Animals were socially housed in groups of 4 and given ad libitum access to food and water. The experimental room was kept on a 12 h light/dark cycle at $22 \pm 2°C$. The animals' weights were monitored and their behaviour assessed in an open field apparatus (620 mm long, 620 mm wide and 150 mm high) both 48 and 24 h prior to commencement of training. The base of the open field box was demarcated into an 8 x 8 grid. The animal was placed in the centre of the box and its locomotion was monitored by counting the number of times its hind legs entered a new square in a 5 min period. All animals assessed and included in the study were behaviourally normal. Other monitored behaviours included rearing, posture and piloerection. Animals were assessed in a quiet room under low-level red light illumination.

Morris Water Maze Training and Tissue Collection

On postnatal day 80, animals were trained in the Morris water maze spatial learning task. Briefly, the water maze apparatus consisted of a large circular pool (150 cm diameter, 80 cm deep) and a hidden platform (11 cm diameter). Both were constructed from black polyvinyl plastic, offering no intra-maze visual cues that may help guide escape behaviour. The platform was submerged 1.5 cm below the water's surface (temperature $26 \pm 1°C$) and positioned 30 cm from the edge of the maze wall. The platform remained in the same location throughout the training session. The experimental room contained several extra-maze visual cues. The rat was lowered into the water facing the wall of the maze (30 cm high) at one of three locations which were alternated with each trial. Trials lasted a maximum of 90 s and the length of time taken for the rat to find the hidden platform was recorded. Rats failing to find the platform within the 90 s were placed on it for 10 s and allowed orientate themselves. The training session consisted of 5 trials with an inter-trial interval of 300 s. Each trained animal was assigned a passive control counterpart animal that spent the same lengths of time swimming in the pool, minus the platform. After training, the rats were dried and placed back into their home cages. They were then killed, without anaesthetic, by cervical dislocation at specific time-points post-training, i.e. 1, 2 or 3 h after commencement of the third trial (n = 4 animals per group, i.e. 24 animals in total). Their brains were quickly dissected out, covered in OCT (optimal cutting temperature compound, Agar Scientific) and snap frozen in CO_2-cooled n-hexane.

Immunofluorescent Double-Labelling of HIF-1α and GFAP

Whole brains were cryosectioned coronally at –3.3 mm with respect to Bregma in order to reveal the dorsal hippocampus [202]. 12 μm sections were adhered to glass slides coated with poly-L-lysine. Sections were fixed in 70% ethanol for 25 min and were then permeabilized using a solution of 0.2% Triton X in PBS for 1 h. Sections were given two 10 min washes in PBS and were then incubated overnight (18 h) with the primary antibodies ab8366 (Abcam) (1:2000 dilution), a mouse IgG antibody that binds to amino acids 329 – 530 of human HIF-1α and G9269 (Sigma), a rabbit IgG antibody which labels GFAP (1:1000 dilution). The primary antibody solution consisted of 1% bovine serum albumin (BSA) and 1% normal goat serum (NGS) in PBS. Following two 10 min washes in PBS, sections were incubated for 3 h with a goat anti-mouse IgG secondary antibody conjugated to FITC (Sigma) (1:1000 dilution) and a goat anti-rabbit IgG secondary antibody labelled with TRITC (Sigma) (1:1000 dilution). The excess secondary antibodies were then washed off and sections were incubated in a solution of Hoechst 33258 (1:500 dilution) for 3 min in order to stain all nuclei. A drop of Citifluor glycerol PBS solution (Agar Scientific) was used to mount each section with a coverslip and the slides were stored in darkness at 4°C.

Image Analysis

All confocal images used for quantitative analysis of HIF-1α expression (12-bit; 1024 x 1024 pixels) were captured using a 40x/0.8W water-dipping lens (Zeiss Achroplan). Three defined neuronal subpopulations of the hippocampus were analysed from hippocampal sections of water maze-trained animals, namely the CA1, CA3 and the apex of the dentate gyrus. The specific areas of the hippocampal neuronal circuit captured were kept consistent between sections (Fig. **4**). Three sections per brain were used for HIF-1α expression analysis. Image analysis was conducted using the software package EBImage (http://www.ebi.ac.uk/~osklyar/EBImage/) for the R statistical programming environment. Nuclear fluorescence intensities of HIF-1α were calculated on a cell-by-cell basis. Briefly, red, green and blue channels were first separated for each image and every pixel within the images (1024 x 1024) was assigned an intensity value between 0 and 1. Using the blue channel as a reference, size and fluorescence intensity thresholds were set in order to select only those pixels likely to represent Hoechst-labelled nuclei. A distance map was then generated for the image which calculates the distance each pixel within the image is from an edge pixel. The watershed segmentation algorithm is then employed which accurately separates clusters of nuclei that are very close together, or touching, into individual cells. Minimum distance between objects and minimum radius criteria are written into the analysis scripts which further refines object separation. Astrocytic HIF-1α fluorescence intensities were quantified by measuring the total green fluorescence within red-labelled astrocytic processes.

ACKNOWLEDGEMENTS

Work featured in this manuscript was funded by a research cluster award from Science Foundation Ireland (03/IN3/B403C and 07/IN.1/B1322).

REFERENCES

[1]   Mons N, Guillou JL, Jaffard R. The role of Ca^{2+}/calmodulin-stimulable adenylyl cyclases as molecular coincidence detectors in memory formation. Cell Mol Life Sci 1999; 55: 525-33.

[2]   Amadio M, Govoni S, Alkon DL, Pascale A. Emerging targets for the pharmacology of learning and memory. Pharmacol Res 2004; 50: 111-22.

[3]   Cajal RY. "La fine structure des centres nerveux." Proceedings of the Royel Society London 1894; 55: 444-68.

[4]   Hebb DO. On human thought. Can J Psychol 1953; 7: 99-110.

[5]   Lechner HA, Squire LR, Byrne JH. 100 years of consolidation—remembering Müller and Pilzecker. Learn Mem 1999; 6: 77-87.

[6]   Morris RGM, Moser EI, Riedel G, *et al.* Elements of a neurobiological theory of the hippocampus: the role of activity-dependent synaptic plasticity in memory. Phil Trans R Soc Lond B 2003; 358: 773-86.

[7]   Lamprecht R, LeDoux J. Structural plasticity and memory. Nat Rev Neurosci 2004; 5: 45-54.

[8]   Bliss TVP, Collingridge GL. A synaptic model of memory: long term potentiation and the hippocampus. Nature 1993; 361: 31-9.

[9]   Malenka RC, Nicoll RA. Long-term potentiation--a decade of progress? Science 1999; 285: 1870-4.

[10]   Geinisman Y. Structural synaptic modifications associated with hippocampal LTP and behavioral learning. Cereb Cortex 2000; 10: 952-62.

[11]   Foley AG, Hartz BP, Gallagher HC, Rønn LC, Berezin V, Bock E, Regan CM. A synthetic peptide ligand of neural cell adhesion molecule (NCAM) IgI domain prevents NCAM internalization and disrupts passive avoidance learning. J Neurochem 2000; 74: 2607-13.

[12]   Conboy L, Murphy KJ, Regan CM. APP expression in the hippocampal dentate gyrus modulates during memory consolidation. J Neurochem 2005; 95: 1677-88.

[13]   Lee TW, Tsang VWK, Birch NP Synaptic plasticity-associated proteases and protease inhibitors in the brain linked to the processing of extracellular matrix and cell adhesion molecules. Neuron Glia Biol 2008; 4: 223-34.

[14]   Marrone DF, Petit TL. The role of synaptic morphology in neural plasticity: structural interactions underlying synaptic power. Brain Res Brain Res Rev 2002; 38: 291-308.

[15]   Regan CM. Memories Are Made of These: From Messengers to Molecules edited by Riedel G, Platt B. Eurekah.com and Kluwer Academic/Plenum Publishers 2004; 564-79.

[16]   O'Malley A, O'Connell C, Regan CM. Ultrastructural analysis reveals avoidance conditioning to induce a transient increase in hippocampal dentate spine density in the 6 hour post-training period of consolidation. Neuroscience 1998; 87: 607-13.

[17]   O'Malley A, O'Connell C, Murphy KJ, Regan CM. Transient spine density increases in the mid-molecular layer of hippocampal dentate gyrus accompany consolidation of a spatial learning task in the rodent. Neuroscience 2000; 99: 229-32.

[18]   Eyre MD, Richter-Levin G, Avital A, Stewart MG. Morphological changes in hippocampal dentate gyrus synapses following spatial learning in rats are transient. Eur J Neurosci 2003; 17: 1973-80.

[19]   Davis HP, Squire LR. Protein synthesis and memory: a review. Psychol Bull 1984; 96: 518-59.

[20]   Bourtchouladze R, Abel T, Berman N, Gordon R, Lapidus K, Kandel ER. Different training procedures recruit either one or two critical periods for contextual memory consolidation, each of which requires protein synthesis and PKA. Learn Mem 1998; 5: 365-74.

[21]   Stork O, Welzl H. Memory formation and the regulation of gene expression. Cell Mol Life Sci 1999; 55: 575-92.

[22]   Quevedo J, Vianna MR, Roesler R, de-Paris F, Izquierdo I, Rose SP. Two time windows of anisomycin-induced amnesia for inhibitory avoidance training in rats: protection from amnesia by pretraining but not pre-exposure to the task apparatus. Learn Mem 1999; 6: 600-7.

[23]   Igaz LM, Vianna MR, Medina JH, Izquierdo I. Two time periods of hippocampal mRNA synthesis are required for memory consolidation of fear-motivated learning. J Neurosci 2002; 22: 6781-9.

[24]   Kandel ER. The molecular biology of memory storage: a dialogue between genes and synapses. Science 2001; 294: 1030-8.

[25]   Lee YS, Bailey CH, Kandel ER, Kaang BK. Transcriptional regulation of long-term memory in the marine snail Aplysia. Mol Brain 2008; 1: 3.

[26]   Abel T, Nguyen PV, Barad M, Deuel TA, Kandel ER, Bourtchouladze R. Genetic demonstration of a role for PKA in the late phase of LTP and in hippocampus-based long-term memory. 1997; Cell 88: 615-26.

[27]   Mansuy IM, Mayford M, Jacob B, Kandel ER, Bach ME. Restricted and regulated overexpression reveals calcineurin as a key component in the transition from short-term to long-term memory. Cell 1998; 92: 39-49.

[28]   Alberini CM. Transcription factors in long-term memory and synaptic plasticity. Physiol Rev 2009; 89: 121-45.

[29]   Freudenthal R, Boccia MM, Acosta GB, *et al.* NF-kappaB transcription factor is required for inhibitory avoidance long-term memory in mice. Eur J Neurosci 2005; 21: 2845-52.

[30]   Cavallaro S, Schreurs BG, Zhao W, D'Agata V, Alkon DL. Gene expression profiles during long-term memory consolidation. Eur J Neurosci 2001; 13: 1809-15.

[31]   Cavallaro S, D'Agata V, Manickam P, Dufour F, Alkon DL. Memory-specific temporal profiles of gene expression in the hippocampus. Proc Natl Acad Sci USA 2002; 99: 16279-84.

[32]   Leil TA, Ossadtchi A, Nichols TE, Leahy RM, Smith DJ. Genes regulated by learning in the hippocampus. J Neurosci Res 2003; 71: 763-8.

[33]   Levenson JM, Choi S, Lee SY, *et al.* A bioinformatics analysis of memory consolidation reveals involvement of the transcription factor c-rel. J Neurosci 2004; 24: 3933-43.

[34]   Keeley MB, Wood MA, Isiegas C, *et al.* Differential transcriptional response to nonassociative and associative components of classical fear conditioning in the amygdala and hippocampus. Learn Mem 2006; 13: 135-42.

[35]   O'Sullivan NC, McGettigan PA, Sheridan GK, *et al.* Temporal change in gene expression in the rat dentate gyrus following passive avoidance learning. J Neurochem 2007; 101: 1085-98.

[36]   Whitney LW, Becker KG, Tresser *et al.* NJ, Caballero-Ramos CI, Munson PJ, Prabhu VV, Trent JM, McFarland HF, Biddison WE. Analysis of gene expression in multiple sclerosis lesions using cDNA microarrays. Ann Neurol 1999; 46: 425-8.

[37]   Mirnics K, Middleton FA, Marquez A, Lewis DA, Levitt P. Molecular characterization of schizophrenia viewed by microarray analysis of gene expression in prefrontal cortex. Neuron 2000; 28: 53-67.

[38]   Haroutunian V, Katsel P, Dracheva S, Stewart DG, Davis KL. Variations in oligodendrocyte-related gene expression across multiple cortical regions: implications for the pathophysiology of schizophrenia. Int J Neuropsychopharmacol 2007; 10: 565-73.

[39]   Haroutunian V, Katsel P, Schmeidler J. Transcriptional vulnerability of brain regions in Alzheimer's disease and dementia. Neurobiol Aging 2009; 30: 561-73.

[40]   Guzowski JF, Setlow B, Wagner EK and McGaugh JL. Experience-dependent gene expression in the rat hippocampus after spatial learning: A comparison of the immediate-early genes *Arc*, c-*fos*, and *zif268*. J Neurosci 2001; 21: 5089-98.

[41]   Colantuoni C, Purcell AE, Bouton CM, Pevsner J. High throughput analysis of gene expression in the human brain. J Neurosci Res 2000; 59: 1-10.

[42]   Sandberg M, Hassett C, Adman ET, Meijer J, Omiecinski CJ. Identification and functional characterization of human soluble epoxide hydrolase genetic polymorphisms. J Biol Chem 2000; 275: 28873-81.

[43]   Aimone JB, Gage FH. Unbiased characterization of high-density oligonucleotide microarrays using probe-level statistics. 2004; J Neurosci Methods 135: 27-33.

[44]   D'Agata V, Cavallaro S. Hippocampal gene expression profiles in passive avoidance conditioning. Eur J Neurosci 2003; 18: 2835-41.

[45]   Kornhauser JM, Greenberg ME. A kinase to remember: dual roles for MAP kinase in long-term memory. Neuron 1997; 18: 839-42.

[46]   Woolf NJ, Zinnerman MD, Johnson GV. Hippocampal microtubule-associated protein-2 alterations with contextual memory. Brain Res. 1999; 821: 241-9.

[47]   Gormezano I, Schneiderman N, Deaux E, Fuentes I. Nictitating membrane: classical conditioning and extinction in the albino rabbit. Science 1962; 138: 33-44.

[48]   Fanselow MS. Contextual fear, gestalt memories, and the hippocampus. Behav Brain Res 2000; 110: 73-81.

[49]   Roth TL, Sweatt JD. Regulation of chromatin structure in memory formation. Curr Opin Neurobiol 2009; 19: 336-42.

[50]   Bertoglio LJ, Joca SR, Guimaraes FS. Further evidence that anxiety and memory are regionally dissociated within the hippocampus. Behav Brain Res 2006; 175: 183-8

[51]   Janz R, Südhof TC, Hammer RE, Unni V, Siegelbaum SA, Bolshakov VY. Essential roles in synaptic plasticity for synaptogyrin I and synaptophysin I. Neuron 1999; 24: 687-700.

[52]   Josephson R, Müller T, Pickel J, *et al.* POU transcription factors control expression of CNS stem cell-specific genes. Development 1998; 125: 3087-100.

[53]   Kee N, Teixeira CM, Wang AH, Frankland PW. Preferential incorporation of adult-generated granule cells into spatial memory networks in the dentate gyrus. Nat Neurosci 2007; 10: 355-62.

[54]   Clelland CD, Choi M, Romberg C, *et al.* A functional role for adult hippocampal neurogenesis in spatial pattern separation. Science 2009; 325: 210-3.

[55]   Smith R. Moving molecules: mRNA trafficking in Mammalian oligodendrocytes and neurons. Neuroscientist 2004; 10: 495-500.

[56]   Barco A, Lopez de Armentia M, Alarcon JM. Synapse-specific stabilization of plasticity processes: the synaptic tagging and capture hypothesis revisited 10 years later. Neurosci Biobehav Rev 2008; 32: 831-51.

[57]   Frey S, Frey JU. 'Synaptic tagging' and 'cross-tagging' and related associative reinforcement processes of functional plasticity as the cellular basis for memory formation. Prog Brain Res 2008; 169: 117-43.

[58]   Meigs TE, Fedor-Chaiken M, Kaplan DD, Brackenbury R, Casey PJ. Galpha12 and Galpha13 negatively regulate the adhesive functions of cadherin. J Biol Chem 2002; 277: 24594-600.

[59]   Kounnas MZ, Moir RD, Rebeck GW, *et al.* LDL receptor-related protein, a multifunctional ApoE receptor, binds secreted beta-amyloid precursor protein and mediates its degradation. Cell 1995; 82: 331-40.

[60]   Cam JA, Zerbinatti CV, Li Y, Bu G. Rapid endocytosis of the low density lipoprotein receptor-related protein modulates cell surface distribution and processing of the beta-amyloid precursor protein. J Biol Chem 2005; 80: 15464-70.

[61]   Banting G, Ponnambalam S. TGN38 and its orthologues: roles in post-TGN vesicle formation and maintenance of TGN morphology. Biochim Biophys Acta 1997; 1355: 209-217.

[62]   Pierce JP, Mayer T, McCarthy JB. Evidence for a satellite secretory pathway in neuronal dendritic spines. Curr Biol 2001;11: 351-5

[63]   McNamara JO 2nd, Grigston JC, VanDongen HM, VanDongen AM. Rapid dendritic transport of TGN38, a putative cargo receptor. Brain Res Mol Brain Res 2004; 127: 68-78.

[64]   Bailey CH, Chen M, Keller F, Kandel ER. Serotonin-mediated endocytosis of apCAM: an early step of learning-related synaptic growth in Aplysia. Science 1992; 256: 645-9.

[65]   Fukazawa Y, Saitoh Y, Ozawa F, Ohta Y, Mizuno K, Inokuchi K. Hippocampal LTP is accompanied by enhanced F-actin content within the dendritic spine that is essential for late LTP maintenance in vivo. Neuron 2003; 38: 447-60.

[66]   Fischer A, Sananbenesi F, Schrick C, Spiess J, Radulovic J. Distinct roles of hippocampal de novo protein synthesis and actin rearrangement in extinction of contextual fear. J Neurosci 2004; 24: 1962-6.

[67]   Smart FM, Halpain S. Regulation of dendritic spine stability. Hippocampus 2000; 10: 542-54.

[68]   Matus A. Actin-based plasticity in dendritic spines. Science 2000; 290: 754-8.

[69]   Arbuzova A, Schmitz AA, Vergères G. Cross-talk unfolded: MARCKS proteins. Biochem J 2002; 362: 1-12.

[70]   McNamara RK, Lenox RH. Distribution of the protein kinase C substrates MARCKS and MRP in the postnatal developing rat brain. J Comp Neurol 1998; 397: 337-56.

[71]   McNamara RK, Stumpo DJ, Morel LM, *et al.* Effect of reduced myristoylated alanine-rich C kinase substrate expression on hippocampal mossy fiber development and spatial learning in mutant mice: transgenic rescue and interactions with gene background. Proc Natl Acad Sci USA 1998; 95: 14517-22.

[72]   Kawano Y, Fukata Y, Oshiro N, *et al.* Phosphorylation of myosin-binding subunit (MBS) of myosin phosphatase by Rho-kinase in vivo. J Cell Biol 1999; 147: 1023-38.

[73]   Maekawa M, Ishizaki T, Boku S, *et al.* Signaling from Rho to the actin cytoskeleton through protein kinases ROCK and LIM-kinase. Science 1999; 285: 895-8.

[74]   Dash PK, Orsi SA, Moody M, Moore AN. A role for hippocampal Rho-ROCK pathway in long-term spatial memory. Biochem Biophys Res Commun 2004; 322: 893-8.

[75]   Pham TH, Clemente JC, Satou K, Ho TB. Computational discovery of transcriptional regulatory rules. Bioinformatics 2005; 21 Suppl 2: ii101-7.

[76]   Huber BR, Bulyk ML. Meta-analysis discovery of tissue-specific DNA sequence motifs from mammalian gene expression data. BMC Bioinformatics 2006; 7: 229.

[77]   Morgan XC, Ni S, Miranker DP, Iyer VR. Predicting combinatorial binding of transcription factors to regulatory elements in the human genome by association rule mining. BMC Bioinformatics 2007; 8: 445.

[78]   Schwartz S, Elnitski L, Li M, *et al.* MultiPipMaker and supporting tools: Alignments and analysis of multiple genomic DNA sequences. Nucleic Acids Res 2003; 31: 3518-24.

[79]   Xie X, Lu J, Kulbokas EJ,*et al.* Systematic discovery of regulatory motifs in human promoters and 3' UTRs by comparison of several mammals. Nature 2005; 434: 338-45.

[80]   Wingender E, Dietze P, Karas H, Knuppel R. TRANSFAC: a database on transcription factors and their DNA binding sites. Nucleic Acids Res 1996; 24: 238-41.

[81]   Sandelin A, Alkema W, Engström P, Wasserman WW, Lenhard B. JASPAR: an open-access database for eukaryotic transcription factor binding profiles. Nucleic Acids Res 2004; 32(Database issue): D91-94.

[82]   Greenberg ME, Siegfried Z, Ziff EB. Mutation of the c-fos gene dyad symmetry element inhibits serum inducibility of transcription in vivo and the nuclear regulatory factor binding in vitro. Mol Cell Biol 1987; 7: 1217-25.

[83]   Rivera VM, Sheng M, Greenberg ME. The inner core of the serum response element mediates both the rapid induction and subsequent repression of c-fos transcription following serum stimulation. Genes Dev 1990; 4: 255-68.

[84]   Miano JM. Serum response factor: toggling between disparate programs of gene expression. J Mol Cell Cardiol 2003; 35: 577-93.

[85]   Moore ML, Wang GL, Belaguli NS, Schwartz RJ, McMillin JB. GATA-4 and serum response factor regulate transcription of the muscle-specific carnitine palmitoyltransferase I beta in rat heart. J Biol Chem 2001; 276: 1026-33.

[86]   Sepulveda JL, Vlahopoulos S, Iyer D, Belaguli N, Schwartz RJ. Combinatorial expression of GATA4, Nkx2-5, and serum response factor directs early cardiac gene activity. J Biol Chem 2002; 277: 25775-82.

[87]   Xu L, Renaud L, Müller JG, et al. Regulation of Ncx1 expression. Identification of regulatory elements mediating cardiac-specific expression and up-regulation. J Biol Chem 2006; 281: 34430-40.

[88]   Arsenian S, Weinhold B, Oelgeschlager M, Ruther U, Nordheim A. Serum response factor is essential for mesoderm formation during mouse embryogenesis. EMBO J 1998; 17: 6289-99.

[89]   Alberti S, Krause SM, Kretz O, et al. Neuronal migration in the murine rostral migratory stream requires serum response factor. Proc Natl Acad Sci USA 2005; 102: 6148-53.

[90]   Knoll B, Kretz O, Fiedler C, et al. Serum response factor controls neuronal circuit assembly in the hippocampus. Nat Neurosci 2006; 9: 195-204.

[91]   Ramanan N, Shen Y, Sarsfield S, et al. SRF mediates activity-induced gene expression and synaptic plasticity but not neuronal viability. Nat Neurosci 2005; 8: 759-67.

[92]   Mann Mj, Dzau VJ. Therapeutic applications of transcription factor decoy oligonucleotides. J Clin Invest 2000; 106: 1071-5.

[93]   Isomura I, Morita A. Regulation of NF-kappaB signaling by decoy oligodeoxynucleotides. Microbiol Immunol 2006; 50: 559-63.

[94]   Dash PK, Orsi SA, Moore AN. Sequestration of serum response factor in the hippocampus impairs long-term spatial memory. J Neurochem 2005; 93: 269-78.

[95]   Dalton S, Treisman R. Characterization of SAP-1, a protein recruited by serum response factor to the c-fos serum response element. 1992; Cell 68:597-612.

[96]   Hipskind RA, Rao VN, Mueller CG, Reddy ES, Nordheim A. Ets-related protein Elk-1 is homologous to the c-fos regulatory factor p62TCF. Nature 1991; 354: 531-4.

[97]   Dalton S, Marais R, Wynne J, Treisman R. Isolation and characterization of SRF accessory proteins. Philos Trans R Soc Lond B Biol Sci 1993; 640: 325-32.

[98]   Giovane A, Pintzas A, Maira SM, Sobieszczuk P, Wasylyk B. Net, a new ets transcription factor that is activated by Ras. Genes Dev 1994; 8: 1502-13.

[99]   Janknecht R, Nordheim A. Elk-1 protein domains required for direct and SRF-assisted DNA-binding. Nucleic Acids Res 1992; 20: 3317-24.

[100]   Yang SH, Whitmarsh AJ, Davis RJ, Sharrocks AD. Differential targeting of MAP kinases to the ETS-domain transcription factor Elk-1. EMBO J 1998; 17: 1740-9.

[101]   Bading H, Ginty DD, Greenberg ME. Regulation of gene expression in hippocampal neurons by distinct calcium signaling pathways. Science 1993; 260: 181-6.

[102]   Xia Z, Dudek H, Miranti CK, Greenberg ME. Calcium influx *via* the NMDA receptor induces immediate early gene transcription by a MAP kinase/ERK-dependent mechanism. J Neurosci 1996; 16: 5425-36.

[103]   Gokce O, Runne H, Kuhn A, Luthi-Carter R. Short-term striatal gene expression responses to brain-derived neurotrophic factor are dependent on MEK and ERK activation. PLoS One 2009; 4: e5292.

[104]   Gineitis D, Treisman R. Differential usage of signal transduction pathways defines two types of serum response factor target gene. J Biol Chem 2001; 276: 24531-9.

[105]   Lanahan A, Worley P. Immediate-early genes and synaptic function. Neurobiol Learn Mem 1998; 70: 37-43.

[106]   Baumgartel K, Tweedie-Cullen RY, Grossmann J, Gehrig P, Livingstone-Zetchej M, Mansuy IM. Changes in the proteome after neuronal zif268 overexpression. J Proteome Res 2009; 8: 3298-316.

[107]   Lavaur J, Bernard F, Frifilieff P, et al. A TAT-DEF-Elk-1 peptide regulates the cytonuclear trafficking of Elk-1 and controls cytoskeleton dynamics. J Neurosci 2007; 27: 14448-58.

[108]   Wang DZ, Li S, Hockemeyer D, et al. Potentiation of serum response factor activity by a family of myocardin-related transcription factors. Proc Natl Acad Sci USA 2002; 99: 14855-60.

[109]   Miralles F, Posern G, Zaromytidou AI, Treisman R. Actin dynamics control SRF activity by regulation of its coactivator MAL. Cell 2003; 113: 329-42.

[110]   Du KL, Ip HS, Li J, et al. Myocardin is a critical serum response factor cofactor in the transcriptional program regulating smooth muscle cell differentiation. Mol Cell Biol 2003; 23: 2425-37.

[111]   Shiota J, Ishikawa M, Sakagami H, Tsuda M, Baraban JM, Tabuchi A. Developmental expression of the SRF co-activator MAL in brain: role in regulating dendritic morphology. J Neurochem 2006; 98: 1778-88.

[112]   Cen B, Selvaraj A, Burgess RC, Hitzler JK, Ma Z, Morris SW, Prywes R. Megakaryoblastic leukemia 1, a potent transcriptional coactivator for serum response factor (SRF), is required for serum induction of SRF target genes. Mol Cell Biol 2003; 23: 6597-608.

[113]   Zaromytidou AI, Miralles F, Treisman R. MAL and ternary complex factor use different mechanisms to contact a common surface on the serum response factor DNA-binding domain. Mol Cell Biol 2006; 26: 4134-48.

[114]   Zhou J, Hu G, Herring BP. Smooth muscle-specific genes are differentially sensitive to inhibition by Elk-1. Mol Cell Biol 2005; 25: 9874-85.

[115]   Hill CS, Wynne J, Treisman R. The Rho family GTPases RhoA, Rac1, and CDC42Hs regulate transcriptional activation by SRF. Cell 1995; 81: 1168-70.

[116]   Carlier MF, Pantaloni D. Control of actin dynamics in cell motility. J Mol Biol 1997; 269: 459-67.

[117]   Guettler S, Vartiainen MK, Miralles F, Larijani B, Treisman R. RPEL motifs link the serum response factor cofactor MAL but not myocardin to Rho signaling *via* actin binding. Mol Cell Bio 2008; 28: 732-42.

[118]   Tabuchi A, Estevez M, Henderson JA, et al. Nuclear translocation of the SRF co-activator MAL in cortical neurons: role of RhoA signalling. J Neurochem 2005; 94: 169-80.

[119]   Blank RS, McQuinn TC, Yin KC, et al. Elements of the smooth muscle alpha-actin promoter required in cis for transcriptional activation in smooth muscle. Evidence for cell type-specific regulation. J Biol Chem 1992; 267: 984-9.

[120]   Yano H, Hayashi K, Momiyama T, Saga H, Haruna M, Sobue K. Transcriptional regulation of the chicken caldesmon gene. Activation of gizzard-type caldesmon promoter requires a CArG box-like motif. J Biol Chem 1995; 270: 23661-6.

[121]   Nakamura M, Nishida W, Mori S, Hiwada K, Hayashi K, Sobue K. Transcriptional activation of beta-tropomyosin mediated by serum response factor and a novel Barx homologue, Barx1b, in smooth muscle cells. J Biol Chem 2001; 276: 18313-20.

[122]   Kalita K, Kharebava G, Zheng JJ, Hetman M. Role of megakaryoblastic acute leukemia-1 in ERK1/2-dependent stimulation of serum response factor-driven transcription by BDNF or increased synaptic activity. J Neurosci 2006; 26: 10020-32.

[123]   O'Sullivan NC, Pickering M, Di Giacomo D, Loscher JS, Murphy KJ. Mkl Transcription Cofactors Regulate Structural Plasticity in Hippocampal Neurons. Cereb Cortex 2010b; 20: 1915-25.

[124]   Miano JM, Long X, Fujiwara K. Serum response factor: master regulator of the actin cytoskeleton and contractile apparatus. Am J Physiol Cell Physiol 2007; 292: C70-81.

[125]   Wickramasinghe SR, Alvania RS, Ramanan N, Wood JN, Mandai K, Ginty DD. Serum response factor mediates NGF-dependent target innervation by embryonic DRG sensory neurons. Neuron 2008; 58: 532-45.

[126]   Stern S, Debre E, Stritt C, Berger J, Posern G, Knoll B. A nuclear actin function regulates neuronal motility by serum response factor-dependent gene transcription. J Neurosci 2009; 29: 5412-8.

[127]   Hai T, Hartman MG. The molecular biology and nomenclature of the activating transcription factor/cAMP responsive element binding family of transcription factors: activating transcription factor proteins and homeostasis. Gene 2001; 273: 1-11.

[128]   Ogryzko VV, Schiltz RL, Russanova V, Howard BH, Nakatani Y. transcriptional coactivators p300 and CBP are histone acetyltransferases. Cell 1996; 87: 953-9.

[129]   Kang H, Sun LD, Atkins CM, Soderling TR, Wilson MA, Tonegawa S. An important role of neural activity-dependent CaMKIV signaling in the consolidation of long-term memory. Cell 2001; 106: 771-83.

[130]   Lau GC, Saha S, Faris R, Russek SJ. Up-regulation of NMDAR1 subunit gene expression in cortical neurons *via* a PKA-dependent pathway. J Neurochem 2004; 88: 564-75.

[131]   Izquierdo LA, Viola H, Barros DM, *et al.* Novelty enhances retrieval: molecular mechanisms involved in rat hippocampus. Eur J Neurosci 2001; 13: 1464-7.

[132]   Athos J, Impey S, Pineda VV, Chen X, Storm DR. Hippocampal CRE-mediated gene expression is required for contextual memory formation. Nat Neurosci 2002; 5: 1119-20.

[133]   Zhang JJ, Okutani F, Inoue S, Kaba H. Activation of the mitogen-activated protein kinase/extracellular signal-regulated kinase signaling pathway leading to cyclic AMP response element-binding protein phosphorylation is required for the long-term facilitation process of aversive olfactory learning in young rats. Neuroscience 2003; 121: 9-16.

[134]   Sindreu CB, Scheiner ZS, Storm DR. Ca^{2+} -stimulated adenylyl cyclases regulate ERK-dependent activation of MSK1 during fear conditioning. Neuron 2007; 53: 79-89.

[135]   Harum KH, Alemi L, Johnston MV. Cognitive impairment in Coffin-Lowry syndrome correlates with reduced RSK2 activation. Neurology 2001; 56: 207-14.

[136]   Nguyen PV, Kandel ER. Brief theta-burst stimulation induces a transcription-dependent late phase of LTP requiring cAMP in area CA1 of the mouse hippocampus. Learn Mem 1997; 4: 230-43.

[137]   Yin JC, Wallach JS, Del Vecchio M, *et al.* Induction of a dominant negative CREB transgene specifically blocks long-term memory in Drosophila. Cell 1994; 79: 49-58.

[138]   Liu RY, Fioravante D, Shah S, Byrne JH. cAMP response element-binding protein 1 feedback loop is necessary for consolidation of long-term synaptic facilitation in Aplysia. J Neurosci 2008; 28: 1970-6.

[139]   Impey S, Smith DM, Obrietan K, Donahue R, Wade C, Storm DR. Stimulation of cAMP response element (CRE)-mediated transcription during contextual learning. Nat Neurosci 1998; 1: 595-601.

[140]   Taubenfeld SM, Wiig KA, Bear MF, Alberini CM. A molecular correlate of memory and amnesia in the hippocampus. Nat Neurosci 1999; 2: 309-10.

[141]   Guzowski JF, McGaugh JL. Antisense oligodeoxynucleotide-mediated disruption of hippocampal cAMP response element binding protein levels impairs consolidation of memory for water maze training. Proc Natl Acad Sci USA. 1997; 94: 2693-8.

[142]   Florian C, Mons N, Roullet P. CREB antisense oligodeoxynucleotide administration into the dorsal hippocampal CA3 region impairs long- but not short-term spatial memory in mice. Learn Mem 2006; 13: 465-72.

[143]   Sheng M, McFadden G, Greenberg ME. Membrane depolarization and calcium induce c-fos transcription *via* phosphorylation of transcription factor CREB. Neuron 1990; 4: 571-82.

[144]   Finkbeiner S, Tavazoie SF, Maloratsky A, Jacobs KM, Harris KM, Greenberg ME. CREB: a major mediator of neuronal neurotrophin responses. Neuron 1997; 19: 1031-47.

[145]   Tao X, Finkbeiner S, Arnold DB, Shaywitz AJ, Greenberg ME. Ca^{2+} influx regulates BDNF transcription by a CREB family transcription factor-dependent mechanism. Neuron 1998; 20: 709-26.

[146]   Cha-Molstad H, Keller DM, Yochum GS, Impey S, Goodman RH. Cell-type-specific binding of the transcription factor CREB to the cAMP-response element. Proc Natl Acad Sci USA 2004; 101: 13572-7.

[147]   Vo N, Kein ME, Varlamova O, *et al.* A cAMP-response element binding protein-induced microRNA regulates neuronal morphogenesis. Proc Natl Acad Sci USA 2005; 102: 16426-31.

[148]   Wayman GA, Davare M, Ando H, *et al.* An activity-regulated microRNA controls dendritic plasticity by down-regulating p250GAP. Proc Natl Acad Sci USA 2008; 105: 9093-8.

[149]   Bartel B, Bartel DP. MicroRNAs: at the root of plant development? Plant Physiol 2003; 132: 709-17.

[150]   Nabel G, Baltimore D. An inducible transcription factor activates expression of human immunodeficiency virus in T cells. Nature 1990; 326: 711-3.

[151]   Karin M. How NF-kappaB is activated: the role of the IkappaB kinase (IKK) complex. Oncogene 1999; 18: 6867-74.

[152]   Kaltschmidt B, Widera D, Kaltschmidt C. Signaling *via* NF-kappaB in the nervous system. Biochim Biophsy Acta 2005; 1745: 287-99.

[153]   Mattson MP, Meffert MK. Roles for NF-kappaB in nerve cell survival, plasticity, and disease. Cell Death Differ 2006; 13: 852-60.

[154]   Romano A, Freudenthal R, Merlo E, Routtenberg A Evolutionarily-conserved role of the NF-kappaB transcription factor in neural plasticity and memory. Eur J Neurosci 2006; 24: 1507-16.

[155]   Meberg PJ, Kinney WR, Valcourt EG, Routtenberg A. Gene expression of the transcription factor NF-κB in hippocampus: regulation by synaptic activity. Mol Brain Res 1996; 38: 179-90.

[156]   Freudenthal, R., Romano, A., Routtenberg, A. Transcription factor NF-kappaB activation after in vivo perforant path LTP in mouse hippocampus. Hippocampus 2004; 14: 677-83.

[157]   Kaltschmidt C, Kaltschmidt B, Baeuerle PA. Brain synapses contain inducible forms of the transcription factor NF-κB. Mech Dev 1993; 43: 135-47.

[158]   Mattson MP, Camandola S. NF-κB in neuronal plasticity and neurodegenerative disorders. J Clin Investig 2001; 107: 247-54.

[159] Wellmann, H., Kaltschmidt, B., Kaltschmidt, C. Retrograde transport of transcription factor NF-κB in living neurons. J Biol Chem 2001; 276: 11821-9.

[160] Meffert MK, Chang JM, Wiltgen BJ, Fanselow MS, Baltimore D. NF-κB functions in synaptic signaling and behavior. Nat Neurosci 2003; 6: 1072-8.

[161] Mikenberg I, Widera D, Kaus A, Kaltschmidt B, Kaltschmidt C. Transcription factor NF-kappaB is transported to the nucleus *via* cytoplasmic dynein/dynactin motor complex in hippocampal neurons. PLoS One 2007; 2: e589.

[162] Eijkenboom M, Van Der Staay FJ. Spatial learning deficits in rats after injection of vincristine into the dorsal hippocampus. Neuroscience 1999; 91: 1299-313.

[163] Nakayama T, Sawada T. (2002) Involvement of microtubule integrity in memory impairment caused by colchicine. Pharmacol Biochem Behav 2002; 71: 119-38.

[164] Kaltschmidt B, Ndiaye D, Korte M, *et al.* NF-κB Regulates Spatial Memory Formation and Synaptic Plasticity through Protein Kinase A/CREB Signaling. Mol Cell Biol 2006; 26: 2936-46.

[165] O'Riordan KJ, Huang IC, Pizzi M, *et al.* Regulation of nuclear factor kappaB in the hippocampus by group I metabotropic glutamate receptors. J Neurosci 2006; 26: 4870-9.

[166] Fox GB, O'Connell AW, Murphy KJ, Regan CM. Memory consolidation induces a transient and time-dependent increase in the frequency of NCAM polysialylated cells in the adult rat hippocampus. J Neurochem 1995; 65: 2796-9.

[167] Sandi C, Merino JJ, Cordero MI, Kruyt N, Murphy KJ, Regan CM. Differential modulation of hippocampal NCAM polysialylation by stressful and traumatic experiences suggests distinct memory encoding mechanisms. Biol Psychiat 2003; 54: 599-607.

[168] Quandt K, Frech K, Karas H, Wingender E, Werner T. MatInd and MatInspector: new fast and versatile tools for detection of consensus matches in nucleotide sequence data. Nucleic Acids Res 1995; 23: 4878-84.

[169] Simpson CS, Morris BJ. Regulation of neuronal cell adhesion molecule expression by NF-kappa B. J Biol Chem 2000; 275: 16879-84.

[170] Grilli M, Ribola M, Alberici A, Valerio A, Memo M, Spano P. Identification and characterization of a kappa B/Rel binding site in the regulatory region of the amyloid precursor protein gene. J Biol Chem 1995; 270: 26774-7.

[171] Mettouchi A, Cabon F, Montreau N, *et al.* The c-Jun-induced transformation process involves complex regulation of tenascin-C expression. Mol Cell Biol 1997; 17: 3202-9.

[172] Wang JH, Manning BJ, Wu QD, Blankson S, Bouchier-Hayes D, Redmond HP. Endotoxin/lipopolysaccharide activates NF-kappa B and enhances tumor cell adhesion and invasion through a beta 1 integrin-dependent mechanism. J Immunol 2003; 170: 795-804.

[173] Richter M, Suau P, Ponte I. Sequence and analysis of the 5' flanking and 5' untranslated regions of the rat N-methyl-D-aspartate receptor 2A gene. Gene 2002; 295: 135-42.

[174] Begni S, Moraschi S, Bignotti S, *et al.* Association between the G1001C polymorphism in the GRIN1 gene promoter region and schizophrenia. Biol Psychiatry 2003; 53: 617-9.

[175] Chiechio S, Copani A, De Petris L, Morales ME, Nicoletti F, Gereau RW 4th. Transcriptional regulation of metabotropic glutamate receptor 2/3 expression by the NF-kappaB pathway in primary dorsal root ganglia neurons: a possible mechanism for the analgesic effect of L-acetylcarnitine. Mol Pain 2006; 2: 20.

[176] O'Sullivan NC, Croydon L, McGettigan PA, Pickering M, Murphy KJ. (2010a) Hippocampal region-specific regulation of NF-kappaB may contribute to learning-associated synaptic reorganisation. Brain Res Bull. 2010a; 81: 385-390.

[177] Aruga J, Mikoshiba K. Identification and characterization of Slitrk, a novel neuronal transmembrane protein family controlling neurite outgrowth. Mol Cell Neurosci 2003; 24: 117-29.

[178] Matsuo N, Terao M, Nabeshima Y, Hoshino M. Roles of STEF/Tiam1, guanine nucleotide exchange factors for Rac1, in regulation of growth cone morphology. Mol Cell Neurosci 2003; 24: 69-81.

[179] Tolias KF, Bikoff JB, Kane CG, Tolias CS, Hu L, Greenberg ME. The Rac1 guanine nucleotide exchange factor Tiam1 mediates EphB receptor-dependent dendritic spine development. Proc Natl Acad Sci USA 2007; 104: 7265-70.

[180] Sheridan GK, Pickering M, Twomey C. Moynagh PN, O'Connor JJ, Murphy KJ. NFκB activity in distinct neural subtypes of the rat hippocampus: Influence of time and GABA antagonism in acute slice preparations Learn Mem 2007; 14: 525-32.

[181] Choi J, Krushel LA, Crossin KL. NF-kappaB activation by N-CAM and cytokines in astrocytes is regulated by multiple protein kinases and redox modulation. Glia 2001; 33: 45-56.

[182] Palin K, Bluthe RM, Verrier D, Tridon V, Dantzer R and Lestage J. Interleukin-1beta mediates the memory impairment associated with a delayed type hypersensitivity response to bacillus Calmette-Guérin in the rat hippocampus. Brain Behav Immun 2004; 18: 223-30.

[183] Boccia M, Freudenthal R, Blake M, *et al.* Activation of hippocampal nuclear factor-κB by retrieval is required for memory reconsolidation. J Neurosci 2007; 27: 13436 - 45.

[184] Henninger N, Feldmann Jr RE, Futterer CD, *et al.* Spatial learning induces predominant downregulation of cytosolic proteins in the rat hippocampus. Genes Brain Behav 2007; 6: 128-40.

[185] Sparkman NL, Buchanan JB, Heyen JRR, Chen J, Beverly JL and Johnson RW. Interleukin-6 facilitates lipopolysaccharide-induced disruption in working memory and expression of other proinflammatory cytokines in hippocampal neuronal cell layers. J Neurosci 2006; 26: 10709-16.

[186] Hu CJ, Wang LY, Chodosh LA, Keith B and Simon MC. Differential roles of hypoxia-inducible factor 1alpha (HIF-1alpha) and HIF-2alpha in hypoxic gene regulation. Mol Cell Biol 2003; 23: 9361-74.

[187] Ke Q and Costa M. Hypoxia-Inducible Factor-1 (HIF-1). Mol Pharmacol 2006; 70: 1469-80.

[188] Zhou J, Schmid T, Brüne B. Tumor necrosis factor-α causes accumulation of a ubiquitinated form of hypoxia inducible factor-1α through a nuclear factor-κB-dependent pathway. Mol Biol Cell 2003; 14: 2216-25.

[189] Jain J, Burgeon E, Badalian TM, Hogan PG, Rao A. A similar DNA-binding motif in NFAT family proteins and the Rel homology region. J Biol Chem 1995; 70: 4138-45.

[190] Okamura H, Garcia-Rodriguez C, Martinson H, Qin J, Virshup DM, Rao A. A conserved docking motif for CK1 binding controls the nuclear localization of NFAT1. Mol Cell Biol 2004; 24: 4184-95.

[191] Rao A, Luo C, Hogan PG. Transcription factors of the NFAT family: regulation and function. Annu Rev Immunol 1997; 15: 707-47.

[192] Hogan PG, Chen L, Nardone J, Rao A. Transcriptional regulation by calcium, calcineurin, and NFAT. Genes Dev 2003; 17: 2205-32.

[193] Graef IA, Mermelstein PG, Stankunas K, *et al.* L-type calcium channels and GSK-3 regulate the activity of NF-ATc4 in hippocampal neurons. Nature 1999; 401: 703-8.

[194] Groth RD, Mermelstein PG. Brain-derived neurotrophic factor activation of NFAT (nuclear factor of activated T-cells)-dependent transcription: a role for the transcription factor NFATc4 in neurotrophin-mediated gene expression. J Neurosci 2003; 23: 8125-34.

[195] Morishita W, Marie H, Malenka RC. Distinct triggering and expression mechanisms underlie LTD of AMPA and NMDA synaptic responses. Nat Neurosci 2005; 8: 1043-50.

[196]   Malleret G, Haditsch U, Genoux D, *et al.* Inducible and reversible enhancement of learning, memory, and long-term potentiation by genetic inhibition of calcineurin. Cell 2001; 104: 675-86.

[197]   Baumgartel K, Genoux D, Welzl H, *et al.* Control of the establishment of aversive memory by calcineurin and Zif268. Nat Neurosci 2008; 11: 572-8.

[198]   Groth RD, Weick JP, Bradley KC, *et al.* D1 dopamine receptor activation of NFAT-mediated striatal gene expression. Eur J Neurosci 2008; 27: 31-42.

[199]   Nguyen T, Lindner R, Tedeschi A, *et al.* NFAT-3 is a transcriptional repressor of the growth-associated protein 43 during neuronal maturation. J Biol Chem 2009; 284: 18816-23.

[200]   Graef IA, Wang F, Charron F, Chen L, Neilson J, Tessier-Lavigne M, Crabtree GR. Neurotrophins and netrins require calcineurin/NFAT signaling to stimulate outgrowth of embryonic axons. Cell 2003; 113: 657-70.

[201]   Schwartz N, Schohl A, Ruthazer ES. Neural activity regulates synaptic properties and dendritic structure in vivo through calcineurin/NFAT signaling. Neuron 2009; 62: 655-69.

[202]   Paxinos G, Watson C. The rat brain in stereotaxic coordinates. 5th ed. 2005; Elsevier Academic Press.

CHAPTER 4

Roles for NF-κB in Regulating Gene Expression in Synaptic Plasticity and Memory

Gary Odero[1], Wanda Snow[1], Kunjumon Vadakkan[1,2] and Benedict C. Albensi[1,3]*

[1]*Division of Neurodegenerative Disorders, St. Boniface General Hospital Research Centre, Winnipeg, MB, Canada;* [2]*Health Sciences Centre, University of Manitoba, Winnipeg, MB, Canada;* [3]*Department of Pharmacology and Therapeutics, University of Manitoba, Winnipeg, MB, Canada*

Abstract: Changing the strength of synaptic connections between neurons is a process by which memory traces are encoded and stored in the nervous system. Evidence to date suggests that long term memory encoding and storage are dependent on mRNA translation and protein synthesis. Studies over the years have identified key signaling molecules involved in processes of protein synthesis in contexts of long term memory. Transcription factors, such as cAMP response element binding (CREB) protein, CCAAT/enhancer binding protein (C/EBP), early growth response (Egr) protein, activator protein 1 (AP-1), and nuclear factor kappa B (NF-κB) have been hypothesized to play roles in memory suggesting that these molecules function as part of a sophisticated response for processes of protein synthesis in long term memory. Previous studies have shown roles for some of these proteins in CNS disorders, where a rapidly growing literature supports the involvement of NF-κB, not only in neurodegenerative conditions, but also in synaptic plasticity and memory.

Keywords: NF-kB, synaptic plasticity, memory, TNF, transcription factor, gene expression, Egr, translation, transcription, neuron, hippocampus, LTP, LTD.

INTRODUCTION

Modern thinking about the formation of memory can be traced to Hebb, who first theorized that the recollection of sensory experiences might be permanently stored by synaptic modification [1]. According to Hebb's postulate, *"When an axon of cell A is near enough to excite a cell B and repeatedly or persistently takes part in firing it, some growth process or metabolic change takes place in one or both cells such that A's efficiency, as one of the cells firing B, is increased."* A paraphrased version of his postulate has also been quoted, *"Neurons that fire together wire together"*. Actual scientific data supporting this postulate came later from a series of ground-breaking papers [2-4] published by a group in Norway that evaluated a novel observation in rabbit brain slices called *frequency potentiation*. The phenomenon of frequency potentiation, which later became known as long-term potentiation (LTP), was described as a long lasting increase in the efficacy of certain brain circuits as a result of brief high frequency electrical stimulation.

Additional work and brainstorming led many investigators to think that changing the strength of synaptic connections between neurons was a mechanism by which memory traces could be encoded and stored in the nervous system. Changing the strength or plasticity of a neural network appeared to be governed by activity-dependent processes that were believed to be responsible for memory formation. To investigate this notion, over 40 years of research effort was devoted to understanding if synaptic plasticity, LTP, and memory are mechanistically related. One unifying idea put forth by Martin *et al.* was that, *"Activity-dependent synaptic plasticity is induced at appropriate synapses during memory formation and is both necessary and sufficient for the information storage*

Address correspondence to Benedict C. Albensi: 351 Tache Ave, St. Boniface Research Centre, R4050. Winnipeg, Manitoba R2H 2A6 Canada; Tel: 204-235-3942; Fax: 204-237-4092; E-mail: BAlbensi@sbrc.ca

Benedict C. Albensi (Ed.)

underlying the type of memory mediated by the brain area in which that plasticity is observed" [5]. This hypothesis, also called the synaptic plasticity and memory (SPM) hypothesis, attempts to "wed" a vast collection of *in vitro* LTP data and studies involving memory in intact animals and humans and is an ongoing synthesis effort even today.

Before a proper discussion of transcription and translational control of SPM can be done, which is the theme of this book chapter, a brief introduction to some of the popular models that are used in SPM studies is warranted. Importantly, both *in vitro* and *in vivo* models have been used in this regard that have provided critical data supporting the SPM hypothesis as defined above, by Martin *et al.* For *in vitro* studies, by far the most popular model has been the hippocampal brain slice procedure utilizing rodents, where a majority of the work has been conducted in the CA1 subfield of the hippocampus [6]. By means of this procedure, the experimental paradigms of LTP and long term depression (LTD) have been used in conjunction to understand underlying mechanisms of SPM. Operationally, LTP and LTD measure potential changes in synaptic potentiation using the excitatory postsynaptic potential (EPSP) as a response index. SPM studies in the intact behaving animal have utilized several models including the electroencephalogram (EEG), *in vivo* electrophysiological recording techniques, and, of course, the Morris water maze, a paradigm assessing spatial memory [7]. Other behavioral tests or instruments include the 8-arm radial maze, elevated plus maze, Y maze, fear conditioning paradigms, the Barnes' maze, and even functional MRI, to name a few [7]. Collectively, these models and paradigms have distinct advantages, but also noteworthy limitations. Therefore, they are usually most effective when used as a group and not in isolation so that a comprehensive assessment of SPM can be made. One thing that these models have in common is that they have been largely used to test hippocampal function in SPM, which, of course, limits what form of memory is being studied. To this end, the models have been widely used to test two general theories of hippocampal function [8]. In the first theory, the hippocampus is thought to be involved in the formation of memories for everyday facts and events where this information can be recalled (a.k.a. declarative memory). The second theory is based on the idea that the hippocampus is involved in spatial memory and the formation of cognitive maps, which are used and recalled during navigation through space.

Given the data from these models, it is now known that there are various types of LTP [3], different forms of memory [9], and several brain areas involved in SPM [10-11]. One question that remains, however, is this: does LTP = memory? Of course such a question is a gross oversimplification of the phenomenon, and any sort of analysis needs to address specific questions, such as what type of LTP and which properties of LTP are relevant to specific forms of memory. In any case, memory processes have been categorized into two large groups involving short term and long term forms [12]. Considerable evidence suggests that both later phase LTP and long term memories are dependent on mRNA translation and protein synthesis [13-15].

To this end, many studies over the years have attempted to identify key signaling molecules involved in processes of protein synthesis in contexts of synaptic plasticity and long term memory. The cyclic AMP (cAMP) responsive element binding protein (CREB) is one such molecule that has been extensively studied for its role in SPM [16-18]. CREB, also a transcription factor, has been found to modulate the transcription of specific genes in contexts of SPM. More recently, other transcription factors, such as CCAAT/enhancer binding protein (C/EBP), early growth response (Egr) protein, activator protein 1 (AP-1), and nuclear factor kappa B (NF-κB) have also been hypothesized to play important roles in SPM [14], suggesting that these molecules function as part of a sophisticated response for processes of protein synthesis in long term SPM.

Transcriptional and Translational Control of Synaptic Plasticity and Memory

Over the last 50 years a large number of studies have been conducted utilizing transcriptional and translational inhibitors, thus implicating and characterizing the requirement for protein synthesis in SPM [14, 19]. One of the first transcriptional inhibitors used in studies of memory by Agranoff *et al.* was actinomycin D [20-21]. During the course of these studies one important observation was made, namely that newly learned information during the memory consolidation phase exists for a temporary period in a labile state. During this labile phase, memory can be disrupted if transcription or translation is inhibited. More recent studies have additionally shown that several waves of protein synthesis occur during learning and memory [22]. In fact, more than one phase of transcription appears necessary for memory formation. This could be one reason why several families of transcription factors (CREB,

C/EBP, Egr, AP-1, and NF-κB) exist and have been found thus far to play a role in SPM. Also of interest is that to date these transcription factor families appear to be dissimilar from those that are involved in the development of the brain and nervous system.

Transcription Factor NF-κB

Before one is able to understand the role of specific transcription factors, such as NF-κB, in SPM, one must first appreciate some basic characteristics of the protein complex. NF-κB belongs to the Rel family of transcription factor proteins [23-24], of which there are five members. These include p50, p52, p65, c-Rel, and RelB. The p65 protein is also known as RelA. These family members function as dimers. However, when NF-κB is at rest in the cytoplasm, it exists in an inactivated state, where the dimer form is bound to IκB inhibitory proteins. Typical IκB inhibitory proteins include IκBα and I κBβ isoforms. Activation of the NF-κB complex initially begins by the degradation of IκB by IκB kinase (IKK). A variety of extracellular signals, such as calcium or cytokines, can stimulate the enzyme IKK. IKK, in turn, phosphorylates the IκB protein, which results in ubiquitination, dissociation of IκBα from NF-κB, and eventual degradation of IκB by the proteosome. Subsequently, the NF-κB dimer translocates to the nucleus, where it induces transcription [25-28]. Typically, the dimer is composed of the p50/p65 subunits, but other combinations are possible. The activated NF-κB dimer binds to the enhancer or promoter region of gene targets and thereby regulates the transcription of genes [29]. The consensus binding site of NF-κB is composed of the GGGRNNYYCC sequence, where R is purine, Y is pyrimidine, and N is any base.

NF-κB Activity in Neurons

NF-κB activity has been previously found in a variety of cell types, especially in immune cells, and in pathological conditions such as cancer and inflammatory responses [30-31]. However, evidence for its activity in neurons is much more recent [32]. During nervous system development, NF-κB is activated in neurons by neurotrophic factors and can induce the expression of genes involved in cell differentiation [33]. Recent studies also show that NF-κB plays a role in cell survival where NF-κB is activated in acute neurodegenerative conditions such as stroke and traumatic injury, as well as in chronic conditions such as Alzheimer's disease [33]. However, some investigators [34] have criticized studies claiming that NF-κB is active in neurons since some of the experiments were conducted with brain cell cultures that varied in their proportions of neurons and glia. Also, there are concerns that NF-κB activation can also occur along a noncanonical pathway, introducing the possibility that the time frame for maximal activation is altered. Nevertheless, a rapidly growing literature supports the involvement of NF-κB, not only in neurodegenerative conditions, but also in synaptic plasticity and memory.

NF-κB in Synaptic Plasticity and Memory

The observation that gene transcription is associated with synaptic activity has led to the hypothesis that synaptic plasticity could result in temporal or long-lasting changes in neuronal structure and function. This theory has been expanded to suggest that second messenger pathways induce activation of a transcription factor, or set of transcription factors, in response to electrical or chemical stimulation. Therefore, this idea links memory-inducing stimuli to gene transcription while also highlighting the importance of transcription factors to the process of memory formation. The notion that NF-κB may play a role in synaptic plasticity and processes underlying memory has arisen, given that signal transduction pathways that lead to the activation of NF-κB, and the subsequent regulation of a number of downstream gene targets by NF-κB, are thought to play important roles in plasticity.

It was Meberg *et al.* who provided the first evidence that linked NF-κB to synaptic plasticity by showing that NF-κB subunits, p50 and p65, increased their mRNA expression in hippocampal granule cells after *in vivo* electrical stimulation of the hippocampal perforant path in rats. Importantly, the stimulation protocol used to upregulate expression also was used to induce LTP in this system [35].

Other behavioral evidence in the crab was presented shortly after this study implicating NF-κB directly in memory processes in *Chasmagnathus* [36-38]. Initial results in the crab showed the enhancement of Rel/κ-B like DNA binding activity in the brain was related to long term habituation formation [37]. In a follow-up study by the same group [36], a learning paradigm was used to assess the potential participation of NF-κB. In this study, the repeated presentation of a danger signal stimulus was used to provoke the decline of an escape response (that was initially elicited). Both intermediate forms and long term forms of memory were evaluated in this model. Interestingly, an enhancement of NF-κB DNA binding activity was found in the long term memory condition, but not after the intermediate term memory condition. Moreover, NF-κB activation was observed after 15 or 30 trials, but not after 5 or 10 trials, consistent with the induction of long term memory, but not intermediate term memory. Merlo *et al.* [38] provided further evidence that NF-κB plays a role in long term memory in *Chasmagnathus* by testing the IκB inhibitor sulfasalazine. Sulfasalazine was used since it was previously found to inhibit IκB kinase (IKK) and thus NF-κB in mammalian models. Using the same learning paradigm they demonstrated that sulfasalazine administered before training resulted in the impairment of long term memory in the crab. In another follow up study in the crab, it was determined for the first time that NF-κB was activated specifically during the memory retrieval process and that this activation was required for memory reconsolidation [39].

Additional experiments were also conducted around the same time that supported the participation of NF-κB in LTP and LTD. For example, Albensi and Mattson [40] showed that the cytokine, tumor necrosis factor alpha (TNFα), which is known to lead to the activation of NF-κB, modulates hippocampal synaptic plasticity. Using TNF double knock out mice, they found, for the first time, that *in vitro* LTD was significantly impaired, and *in vitro* LTP was slightly reduced in CA1 hippocampus. In addition, they found LTD and LTP were significantly impaired in the hippocampus when NF-κB decoy DNA (blocks NF-κB from binding to NF-κB binding sites on DNA) was applied to wild type mouse hippocampal brain slices (Fig. **1**). In this experiment, scrambled DNA or 50 µM κB decoy DNA were administered to CA1 hippocampal slices and 100 or 1 Hz stimulation applied to the Schaffer collateral axons. Slices with κB decoy DNA did not show LTD responses in contrast with slices treated with vehicle that showed normal LTD. In addition, when compared to the response in control slices, the magnitude of LTP following 100 Hz stimulation was significantly attenuated in slices preincubated with κB decoy DNA for approximately 5 hours. Treatment of hippocampal slices with scrambled control double stranded oligodeoxynucleotide had no effect on the degree of LTP, demonstrating the specificity of the effect exhibited with κB decoy DNA. Interestingly, incubation of slices with NF-κB decoy DNA for 2 hours or less had no significant effect on LTP or LTD, which is in line with the idea that the involvement of a transcription-dependent process is occurring in this context at later time points. Freudenthal *et al.* [41] similarly performed LTP and LTD experiments, but in an *in vivo* setting, and found that NF-κB was activated following perforant path stimulation for LTP in mouse hippocampus. These results suggested that NF-κB activation was important for the gene expression required in the maintenance of LTP. In addition, they provided immunostaining evidence (in post-stimulated tissue) of an increase in activated p65 in most subregions of the hippocampus. Collectively, these data indicated important roles for NF-κB signaling pathways in a context of hippocampal synaptic plasticity.

In a study by Kassed *et al.* [42] using a NF-κB p50 knock out mouse in an active avoidance step-up test, the absence of the p50 subunit was found to result in difficulties in task acquisition. In the active avoidance test, animals are placed in a setting they explored previously where they must learn to find a platform in order to escape a modestly painful foot-shock stimulus. These results showed that the absence of p50 not only negatively modulated learning and memory in this paradigm, but also altered hippocampal responsiveness to brain injury after a chemical-induced lesion. Freudenthal *et al.* [43] also used an inhibitory avoidance paradigm and tested NF-κB inhibitors (sulfasalazine and κB decoy) and drugs that do not inhibit the NF-κB pathway (indomethacin). Here they found that intracerebral ventricular injections of sulfasalazine and κB decoy impaired long term memory in this paradigm, whereas indomethacin had no effect on memory. These results, for the first time, supported the notion that NF-κB activity is necessary in the hippocampus during memory consolidation.

Also in 2002 Yeh *et al.* [44] showed for the first time in intact animals that NF-κB was activated in the amygdala following the fear-potentiated startle response, which demonstrated a requirement for NF-κB activation during the consolidation of fear memory. Other studies by Yeh *et al.* [45] also in intact animals further demonstrated a role for NF-κB in memory. In this study, utilizing a fear conditioning paradigm, rats were pretreated with histone deacetylase (HDAC) inhibitors that resulted in enhanced long term fear memory, but not short term fear memory. HDAC inhibitors are known to prolong the nuclear expression of p65 followed by increased DNA binding activity of its acetylated form. The enhancement was reversed with applications of κB decoy DNA.

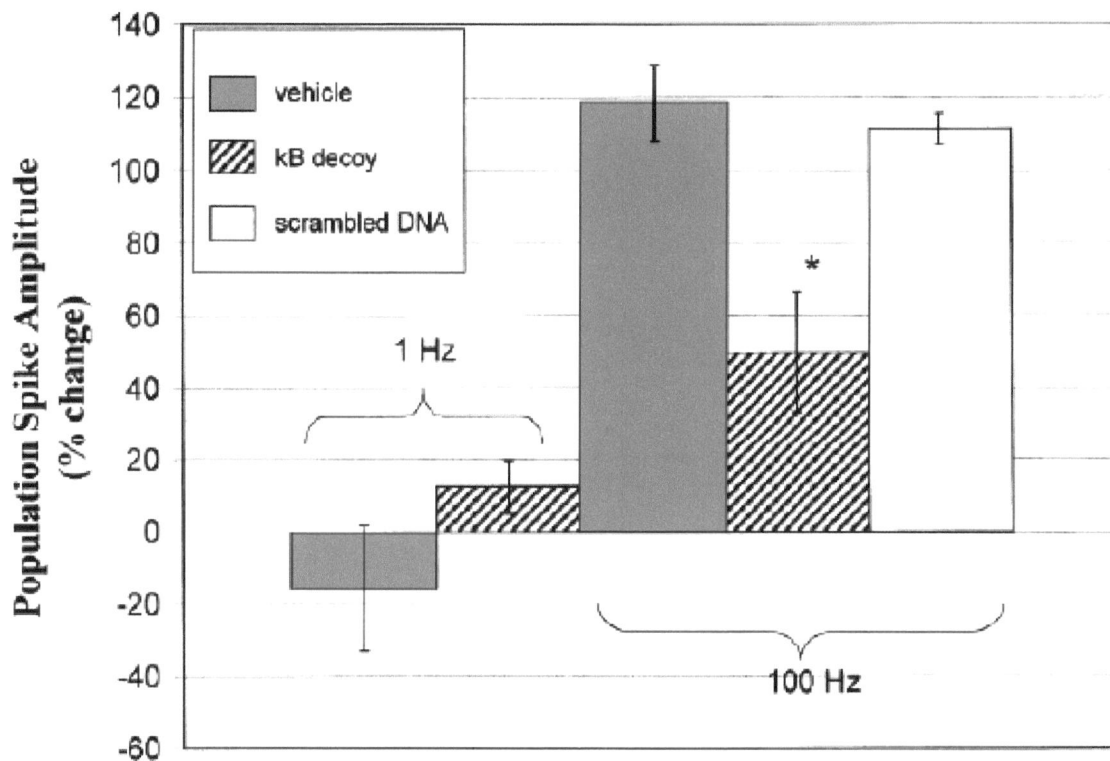

Fig. (1). Shown above are the effects of NF-κB decoy DNA treatment on LTD and LTP responses in mouse hippocampal slices. Slices from wild-type C57BL/6 mice were pretreated for ~5 h with vehicle (saline), 50 μM NF-κB decoy DNA, or 50 μM scrambled DNA, and then stimulated at either 1 or 100 Hz as indicated. The data clearly show that NF-κB decoy DNA, an inhibitor of NF-κB activity, significantly decreased the LTP response (population spike amplitude) following 100 Hz stimulation and eliminated the LTD response following 1 Hz stimulation. Used with permission and modified from Albensi and Mattson, *Synapse* 35:151-159, 2000.

Bioinformatic analyses of memory consolidation (contextual fear memory) have also been conducted in NF-κB c-Rel knock out mice showing that NF-κB c-Rel is necessary for hippocampal dependent long term memory formation [46]. In this investigation, data also suggested that multiple waves of transcription are possibly involved in the consolidation of long term memory since levels of various transcription factors (egs., Egr-1, c-fos, TBr-1, bHLH, MREBP2, etc.) were regulated at discrete times during the course of this study.

Other knock out studies evaluating NF-κB function have included NF-κB p65 knock out mice [47]. Normally, deleting the p65 subunit gene is embryonically lethal due to hepatic failure. So in order to bypass this problem, TNFR-/-p65-/- double knock out mice were used where mutation of the TNF receptor 1 rescues lethality. In this study, these mice demonstrated a selective learning deficit in the spatial version of the 8-arm radial maze. These data also showed a new calcium/calmodulin-dependent kinase II (CaMKII)-dependent mechanism of transcriptional activation since NF-κB activation by calcium required CaMKII.

Studies utilizing mice that conditionally express a potent and nondegradable NF-κB/Rel inhibitor, termed IκBα superrepressor (IκBα-SR), where the superrepressor is robustly expressed in inhibitory GABAergic interneurons, have also been recently conducted [48]. NF-κB activation, as described above, is known to result in p50/p65 dimers rapidly translocating to the nucleus where they engage cognate κB enhancer elements (gene targets). NF-κB activation is initiated once the IκB inhibitory subunit is acted upon by IKK. In one of these gene targets, the IκB gene is contacted by activated NF-κB, which elicits an auto-inhibitory feedback loop. The creation of an IκBα superrepressor, obtained by substitution of 2 key serine phosphor-acceptor sites in IκBα with alanine amino acids (S32A/S36A), serves as a tool for investigating NF-κB activity in regulating inhibitory and excitatory function in SPM. In this study, mice expressing IκBα-□SR showed increased *in vitro* LTP and improved performance in the Morris water maze and the radial arm maze consistent with the idea that repressing the IκBα gene, which codes for the inhibitory subunit of NF-κB, will reduce auto-inhibitory feedback associated with NF-κB activation.

Gene Targets of NF-κB

NF-κB functions as a transcription factor, where the gene targets that mediate the effects of NF-κB on synaptic function in the nervous system are still emerging. Past studies in other areas, such as cancer biology [49-50], have determined that NF-κB is a stress-regulated transcription factor that has a pivotal role in the control of inflammatory and innate immune responses. For example, NF-κB activation quickly induces the transcription of genes encoding cell adhesion molecules, cytokines, cytokine receptors, and enzymes that generate inflammatory mediators. More recent work also shows that NF-κB is an important regulator of genes involved in cell survival and inflammation in cerebral hypoxia-ischemia in neurons [51]. Yet other studies have implicated NF-κB and specific gene targets in seizure activity [52-53]. Interestingly, some studies suggest that NF-κB may induce genes encoding both death proteins and survival proteins [54] that play a role in brain injury; however how the expression of these gene targets contributing to neuroprotection or neurodegeneration is still being actively investigated. For example, several labs have demonstrated that activation of NF-κB can modulate the expression and/or regulation of specific subunits of *N*-methyl-D-aspartate (NMDA) [55] and 2-amino-3-hydroxy-5-methylisoxazole-4-propionic acid (AMPA) receptors [56-58], which are important receptors in brain injury processes and processes of memory formation. Other NF-κB target genes that have been found in neurons include amyloid precursor protein (APP), CaMKII δ, brain-derived neurotrophic factor, μ-opioid receptors, neural cell adhesion molecule, and nitric oxide synthetase [59], pointing to roles in Alzheimer's disease, neuronal development, and other aspects of acute and chronic brain injury.

More specific to cognition, several studies have highlighted the critical role of immediate early genes (IEGs) to memory formation by showing that inhibiting the synthesis of new proteins following behavioral training interferes with the consolidation of memories [60-62]. In particular, some studies have linked NF-κB activation with the transcription of IEGs. Work with IEGs and SPM led to the idea that there are early and late phases of LTP, which are dependent on unique cellular and molecular mechanisms.

The early phase of LTP, lasting approximately 1-3 hours, is under the control of second messengers and protein kinases [63] and does not depend on protein and RNA synthesis. Because IEGs initiate an immediate cascade of signaling events in addition to interacting with downstream synaptic genes or proteins, it has been hypothesized that activation of key IEGs may signal a critical biochemical event bringing about the mechanisms that may underlie the maintenance of synaptic plasticity and memory in general. Of the known IEGs, NF-κB has been shown to regulate the transcription of c-fos, zif268 and arg3.1.

One of the earliest IEGs thought to be up-regulated following LTP induction is zif268, also known as Egr-1 or Krox-24 [64-65]. Additional studies have supported this finding showing that stimulation of inhibitory circuits known to block LTP also resulted in a blockage in IEG induction [64]. Based on this evidence, it has been hypothesized that the activation of zif268 and the induction of LTP are regulated by the same mechanisms. In addition, knockout studies have revealed impairments in behavioral capacity, induction of LTP and proper expression of late LTP [66], indicating that for full expression of LTP to occur, normal expression of zif268 is required. Similar evidence has been provided for another IEG, ie., c-fos. Genetic evidence has indicated hippocampal-dependent spatial memory deficits occurs when c-fos knockout mice are tested in the Morris water maze [67]. Early studies of arc/arg3.1 have also shown that knockdown of this IEG results in normal induction, but adversely affected maintenance of LTP, essentially eliminating the late phase of LTP entirely [68]. Behaviorally, these mice displayed memory deficits in spatial learning, fear conditioning, and object recognition [69]. These data collectively argue for a role of arc/arg3.1 in the maintenance phase of LTP that is thought to be essential for the storage of long-term memories. While IEGs appear to modulate genes essential for the induction of LTP, and prime neurons for the biochemical events required for long-lasting memory storage, late LTP requires activation of a different set of genes.

Late LTP consists of a longer-lasting phase of usually 3-6 hours, is characterized by the requirement of new protein and RNA synthesis, and is mediated primarily by cAMP and CREB [70-74]. The short-term enhancement of synaptic strength occurs by means of post-translational modification of preexisting proteins and is mediated by both cAMP-dependent protein kinase (PKA) and protein kinase C (PKC) [75]. The long-term process requires new protein synthesis and cAMP-mediated gene expression through the activation of CREB-1 and the relief from repression of CREB-2 [76-77] and leads to the growth of new synaptic connections critical for synaptic plasticity [78-79].

Interactions Between NF-κB and CREB in Synaptic Plasticity and Memory

Of particular note is the fact that NF-κB has been known to functionally modulate the expression of CREB, further substantiating a critical role for NF-κB in various phases necessary for memory processes [80]. The CREB signaling pathway has previously been shown to play a key role in memory and learning processes in *Drosophila*, *Aplysia* and rodents [11, 81-84]. The first evidence linking CREB to synaptic plasticity came from the observation that stimulation that induces late LTP (L-LTP) in the hippocampus also induced CRE-mediated gene expression [85] and CREB phosphorylation [86]. Consistent with this, induction of late LTP (L-LTP) in hippocampal slices produced either by tetanic stimulation or application of a membrane-permeable cAMP analog could be blocked by inhibitors of protein synthesis or transcriptional blockers [70, 72, 74]. Similarly, L-LTP-induced gene expression could be blocked by PKA inhibitors or L-type Ca^{2+} channel blockers [72], again implicating the CREB pathway as a convergence point for both cAMP and Ca^{2+} signaling.

The cAMP-PKA signaling mechanism has received much attention in learning-related gene expression because inhibition of PKA in transgenic mice overexpressing the mutant regulatory subunit of PKA specifically impairs late-phase, transcription-dependent LTP and the animals' performances of learning tasks [87]. In addition, several learning mutants in Drosophila have mutations in genes involved in cAMP metabolism [81], and, in the excitable cell lines PC12 and AtT20, inhibition of PKA by pharmacological or genetic means blocks calcium-influx induced gene expression [88]. Thus, a large number of studies appear to indicate that cAMP functions as the messenger that links synaptic activity to genetic responses, as well as stressing the importance of activation of such a pathway to the maintenance of LTP required for memory storage.

Recent data also now appear to link NF-κB activation, PKA, and CREB signaling. In a study by Kaltschmidt *et al.* [80] NF-κB was found to regulate spatial memory formation *via* a PKA/CREB signaling pathway. Results from this study showed that cyclic AMP (cAMP)-dependent PKA (PKAcatα) is a new NF-κB target gene and that its downregulation in neuron-targeted NF-κB mutant mice affects the CREB pathway. For example, NF-κB inhibition in neurons resulted in a drastic downregulation of PKA gene expression and showed a decrease in forskolin-induced CREB phosphorylation. In spite of these interesting results in the PKA/CREB pathway, the role of NF-κB in SPM may still involve additional unidentified signaling pathways that wait future investigation.

Chromatin Modification in Regulating Transcription for Memory

One aspect that is sometimes forgotten is that transcription is not occurring on naked DNA, but is in fact occurring in an environment of chromatin [89]. Chromatin is a complex of histone proteins and DNA, responsible for condensing and organizing DNA in the nucleus. The repeating unit of chromatin is the nucleosome. The nucleosome is made of a histone octamer that interacts with ~147 base pairs of genomic DNA for each nucleosome. The histone octamer itself contains pairs of histones H2A, H2B, H3, and H4. During the process of gene initiation, chromatin modifying enzymes come into play in order to modify the chromatin structure. The regulation of chromatin structure is rapidly emerging as an important area for the field of memory research. Part of this regulation process involves two groups of enzymes that control levels of histone acetylation, which are the histone acetyltransferases (HATs) and the histone deacetylases (HDACs). One function of histone acetylation carried out by HATs is to relax chromatin structure, thus permitting greater access to the DNA by transcription factors. As of this writing, several HATs have been implicated in memory formation with most of the work thus far conducted on CREB-binding protein (CBP) [89]. CBP is, of course, a coactivator recruited by CREB during CREB nuclear signaling. On the other hand, HDAC activity and histone deacetylation have been associated with transcriptional repression. Therefore, HATs and HDACs work together by balancing each other to modulate gene expression.

Given that decreases in histone acetylation and administration of HDACs lead to memory impairments, it was quickly realized that HDAC inhibitors may enhance SPM. And as mentioned earlier, in a study by Yeh *et al.* [45], the administration of HDAC inhibitors resulted in enhanced long term memory in a fear conditioning paradigm. Other recent data has also shown that IκB kinase regulates chromatin structure during the reconsolidation of conditioned fear memories, further implicating the regulation of chromatin structure in memory processes [46]. In this study, a contextual fear conditioning protocol was also used to assess memory reconsolidation. Here it was found that NF-κB signaling activity was upregulated and was associated with increases in histone H3 phosphorylation and acetylation triggered after the retrieval of contextual conditioned fear memory. It was also found that inhibition of NF-κB signaling at the level of IKKα protein kinase blocked both histone H3 posttranslational modifications in hippocampus after retrieval and blocked contextual conditioned fear memory reconsolidation. Importantly, this data links IKKα protein kinase and its regulation of histone H3 modifications to memory reconsolidation in an *in vivo* setting.

Alternative Models for Long Term Memory

The above discussion spent considerable time attempting to convince the reader that long term memories are associated with activity dependent mRNA translation and transcriptional activity leading to protein synthesis. This theory is also known as the protein synthesis theory of long term memory. However, not all investigators hold this view. An alternative model is proposed by some based on the fact that there are instances of memory storage that occur in the absence of protein synthesis [90]. Central to this view is the idea that post-translational protein modification (PTM) is the only mechanism required for long term memory. PTM is not a new idea in and of itself, and in fact PTM is the leading candidate mechanism proposed for short term plasticity. Another postulate of the PTM view is that PTMs are updated and maintained by endogenous activity, which functions as a continuous rehearsal mechanism. Future studies are therefore needed to reconcile any outstanding difficulties with the protein synthesis model or to find additional evidence to support the PTM model so that a unified theory can be put forth.

Drug Therapy for Memory Improvement

Changes in plasticity occurring at synapses during learning and memory require protein molecules and are regulated by gene transcription in the nucleus, as discussed above. To accomplish this requirement, signaling from the synaptic cleft to the nucleus needs to take place. This process, of course, involves a highly complex chain of molecular events. Pathological changes associated with most memory disorders leads to disruptions at various stages in this process. For example, pathophysiological changes such as excitotoxicity, calcium overload, degeneration of cholinergic neurons, deposition of amyloid plaques, and neurofibrillary tangles, are key pathological processes that have been found to contribute to memory dysfunction [91-94]. Current therapeutic interventions focus either on halting or reversing these pathological and molecular changes. However, the noticeably long distance from the synapse to the nucleus may make the regulation of the nuclear events in response pathological changes more complex than currently appreciated. In this context NF-κB, a molecule that is present in the synapses as well as in the nucleus, may serve to circumvent the complexity of this process. If this is the case, then targeting the regulation of the NF-κB molecule may produce therapeutic advantages in memory disorders.

Examination of the mode of action of currently used pharmaceutical agents may provide some insight about the response of the whole nervous system to systemic use of pharmaceutical agents to improve memory. Two of the currently used therapeutic agents for Alzheimer's disease include the NMDA antagonist, memantine, and the acetylcholinesterase inhibitor, donepezil. They act by different mechanisms to achieve improvements in memory; however, to date the improvements have been modest at best. The therapeutic approach using memantine evolved, since in previous studies it was found that blocking NMDA receptors from too much glutamate resulted in neuroprotection in cell models. Excessive glutamate can damage neurons and therefore lead to memory loss. Memantine is a moderate affinity, noncompetitive and uncompetitive NMDA receptor antagonist with strong voltage-dependency and fast kinetics possibly acting by reducing the excitotoxicity of the glutamatergic neurons

[95]. In retrospect, it is apparent that it is excitotoxicity that produces cell loss in dementia and not a glutamate deficiency. A second example is donepezil, which is a centrally acting reversible acetylcholinesterase inhibitor which functions by increasing acetylcholine at cholinergic synapses. Its main therapeutic use is in the treatment of Alzheimer's disease where it is used to increase cortical acetylcholine levels. Positive results using this approach support the notion that the neurotransmitter acetylcholine is essential for memory mechanisms, and its deficiency can lead to memory loss. These two agents demonstrate that pathological conditions that lead to memory loss vary with different neurotransmitters. The above results should be viewed as the effects of pharmacological agents during systemic applications. The questions that arise from the above examples are: 1) what type of changes in NF-κB signaling will lead to memory loss? 2) what effect does systemic application of medications have in modulating NF-κB signaling? Addressing these questions is crucial in initiating new drug therapy.

At the behavioral level, several animal experiments have showed links between NF-κB and memory. For example, the involvement of NF-κB in long-term retention of fear memory [44-45], in inhibitory avoidance memory [96], and spatial long-term memory [97] has been demonstrated. Either RelA or c-Rel and p50 NF-κB factors were found to be involved in mechanisms of cognition [42, 46, 59]. In forebrain neuronal conditional NF-κB-deficient mice, the loss of NF-κB impairs synaptic transmission, spatial memory formation, and plasticity [80].

At the cellular level, NF-κB is known for its roles ranging from neuronal survival [98-99] to memory mechanisms [100]. Among five different members of NF-κB, only RelA, c-Rel, and RelB are directly able to activate the transcription of target genes [99]. The active transcription factor is a dimer that is formed from the above three forms. It is known that memory loss can result if there are defects in new granule neuron formation. NF-κB signaling has also been shown to regulate cell differentiation and survival [101-104]. It is also known that alterations in shape of dendritic spines are an enduring structural correlate of synaptic plasticity [105]. In this context, NF-κB signaling that regulates the growth of neuronal processes of maturing neurons [102-104, 106] may be essential for memory formation and maintenance.

Most importantly, NF-κB p50 was shown to have a role in promoting neurogenesis [107]. Therefore, it is possible that inducing NF-κB in conditions that are known to impair memory due to reduced neurogenesis in the granule neuron layer may reverse the impairment in memory. For example, one of the characteristic features of depression is pseudo-dementia that is associated with memory loss. Antidepressants can reverse depression by inducing neurogenesis [108-109]. The presence of some forms of NF-κB both in the synapses and in the nucleus as a transcription factor could provide another opportunity to target this molecule. If it can be demonstrated that there is an alteration in NF-κB protein in particular disease conditions, then modulating the availability of this molecule or activating its gene transcription may provide another opportunity to improve memory loss.

However, given the complexity of NF-κB signaling, one important challenge remaining is determining if we should enhance or inhibit the activation NF-κB. This is an important point considering past studies have shown dichotomous signaling pathways associated with NF-κB activity, where activity along one route results in neuroprotection, and activity along another path results in cell death. Inflammatory processes are certainly central to both cancer and Alzheimer's disease, but the end result is quite different in each condition since cancer results in unregulated cell growth, and Alzheimer's disease is associated with cell loss. Therefore, future studies most certainly will focus on selected targets and "optimal activation" levels during critical stages of specific disease pathologies to address this issue.

To date, commercial efforts in the pharmaceutical industry to develop modulators of the NF-κB pathway have largely focused on selective IKK-β inhibitors. These efforts are mostly directed towards limiting inflammatory processes. This includes agents such as PS-1145, SPC-839, and BMS-345541, among others [110]. Targets for inhibiting the NF-κB pathway have been categorized into seven different groups [110]. These include agents that inhibit NF-κB protein expression and binding to DNA (RelA antisense, NF-κB decoy and RNAi); agents that interfere with the formation of the IKK complex (NBD peptide, which corresponds to the IKK-γ-binding domain of IKK-β); compounds that block the IKK-β activation process (TIRAP peptide that corresponds to the Toll-interleukin-1 receptor (TIR) domain-containing adapter domain (TIRAP)); Hsp90 inhibitor: Hsp90 stabilizes RIP proteins that are the components of the tumor necrosis factor alpha (TNF-α) receptor signaling complex and antioxidants); molecules that inhibit IKK-β kinase activity; proteasome inhibitors; agents that are inhibitors of NF-κB nuclear translocation (IκB α super-repressor and SN50: a peptide consisting of the nuclear localization sequence (NLS) of the p50 NF-κB subunit); and inhibitors of NF-κB transcriptional activity (glucocorticoids).

In addition, several non-steroidal anti-inflammatory drugs (NSAIDs) have been tested for their ability to inhibit NF-κB activation. Results thus far indicate that these compounds are competitive inhibitors of the ATP-binding site of IKKβ, and so they impair the phosphorylation of IκB and the activation of NF-κB [111]. Some investigators have also tested immunomodulatory drugs (eg., thalidomide) for their anti-inflammatory and immunosuppressive effects and their potential ability to inhibit NF-κB activation by modulating cytokines, such as TNFα and interleukin IL-6 [112]. Additional approaches have also been attempted that include using small interfering RNA (siRNA) that modulate the expression of IKK proteins through RNA interference and the use of cell permeable peptides, which contain the IKK-γ-binding motif and compete with IKK-α and IKK-β for binding to IKK-γ, therefore blocking the activation of the NF-κB pathway [110].

FUTURE DIRECTIONS

Given the currently recognized, but poorly understood gene targets of NF-κB, the gene targets that have yet to be identified, and the potential interactions of NF-κB with other transcriptional families, it should be obvious that there is still much work to be done in this area. One specific research direction that our laboratory is starting to investigate involves links between NF-κB activation and other transcriptional families. For example, as mentioned above, NF-κB belongs to one of five transcriptional families (CREB, C/EBP, AP-1, Egr, and NF-κB) that have been implicated in SPM. To date, our preliminary findings (unpublished observations soon to be in press) show novel links between NF-κB activation and expression of Egr protein. Other data reviewed above also showed that NF-κB activation was linked to PKA/CREB signaling, suggesting that NF-κB signaling may be a higher order or meta-regulator of many transcriptional processes serving memory formation. However, many questions still remain. What are the differences and similarities among the 5 transcriptional families? Perhaps transcriptional families have characteristic functions or even antagonistic functions in SPM. Are there master transcriptional regulators that influence several aspects of transcription in SPM? Is NF-κB one of these regulators? Do several families of transcriptional regulators exist in SPM in order to generate multiphasic waves of activity? If so, what function do multiple waves of transcriptional activity in SPM serve? What roles still need to be understood with regard to transcriptional activators versus transcriptional repressors? What role does NF-κB and/or chromatin play in the transcriptional activation versus repression of activity? What is the functional significance of Egr once it is activated by NF-κB? This is one question our lab is trying to address.

Recent work with Egr family members suggest that these proteins (Egr-1, Egr-2, Egr-3, Egr-4) may have different, and in some cases, antagonistic functions in the brain with respect to cognitive processes [113]. Previous studies have shown that all 4 family members are highly expressed in CA1-CA3 pyramidal cells [114]. To date, however, Egr-1 is the most studied of all the members. In particular, it has been shown that Egr-1 is specially required for the consolidation of long term memory, while Egr-3 appears to be necessary for short term memory [113]. Interestingly, Egr-2, on the contrary, might serve some yet unknown inhibitory function in cognition based on recent studies in Egr-2 knock out mice [113]. Egr-2 has been implicated previously in peripheral nerve myelination, but overall very little is known about Egr-2 or Egr-4 function [113]. Future studies will need to investigate not only examples from each transcriptional family, but also how specific transcription factors function and potentially interact with one another to regulate SPM.

Moreover, much work is still needed in order to clarify how NF-κB signaling might be effectively targeted for human drug therapy in memory disorders. Importantly, one has to first ask the question if NF-κB signaling is too broad in scope to serve well as a target for treating the dysfunction of memory. If work in this area does continue, what is the most reasonable agent to use, and where in the NF-κB pathway should it be used for treating memory disorders or for enhancing memory? And do we inhibit or enhance NF-κB activation in order to improve memory? The best answer right now might be simply to say, *it depends*. In studies involving potential cancer therapies, it appears that most approaches have involved the inhibition of NF-κB signaling. However, in all SPM studies to date, it appears that NF-κB is a requirement for SPM, and so enhancing its activation might be the correct approach. This also leads to the potential concern, though, that if too much NF-κB signaling is triggered, it might lead to enhanced tumor proliferation. As one can see, many critical questions are still unanswered, but the role of NF-κB signaling in SPM will continue to be a major research focus for years to come.

ACKNOWLEDGEMENTS

Research in my laboratory (B.C.A.) was supported by the St. Boniface General Hospital Research Foundation, the Everett Endowment, the Dept. of Pharmacology - University of Manitoba, Centre on Aging - University of Manitoba, the Manitoba Health Research Council (MHRC), the National Sciences and Engineering Research Council of Canada (NSERC), the Scottish Rite Charitable Foundation of Canada, the Alzheimer's Society of Canada, the Manitoba Medical Service Foundation (MMSF), and the Canadian Institutes of Health Research (CIHR).

REFERENCES

[1]   Hebb DO. Oranization of Behavior. Wiley: New York, 1949.
[2]   Anderson P, Lomo T. Mode of activation of hippocampal pyramidal cells by excitatory synapses on dendrites. Exp Brain Res 1966; 2: 247-60.
[3]   Bliss TV, Lomo T. Long-lasting potentiation of synaptic transmission in the dentate area of the anaesthetized rabbit following stimulation of the perforant path. J Physiol 1973; 232: 331-56.
[4]   Lomo T. Frequency potentiation of excitatory synaptic activity in the dentate area of the hippocamal formation. Acta Physiologica Scandinavica 1966; Suppl 277: 128.
[5]   Martin SJ, Grimwood PD, Morris RG. Synaptic plasticity and memory: an evaluation of the hypothesis. Annu Rev Neurosci 2000; 23: 649-711.
[6]   Bliss T, Collinridge G, Morris R, Synaptic plasticity in the hippocampus. In The Hippocampus Book, Anderson P Morris R Amaral D Bliss T O'Keefe J Eds. Oxford University Press: New York, 2007; pp 343-474.
[7]   O'Keefe J, Hippocampal neurophysiology in the behaving animal. In The Hippocampus Book, Oxford University Press: New York, 2007; pp 474-548.
[8]   Morris R. Theories of hippocampal function in The Hippocampus Book. Oxford Unviersity Press: New York, 2007.
[9]   Squire LR, Knowlton B, Musen G. The structure and organization of memory. Annu Rev Psychol 1993; 44: 453-95.
[10]   Zola-Morgan S, Squire LR, Amaral DG. Human amnesia and the medial temporal region: enduring memory impairment following a bilateral lesion limited to field CA1 of the hippocampus. J Neurosci 1986; 6: 2950-67.
[11]   Kandel ER. The biology of memory: a forty-year perspective. J Neurosci 2009; 29: 12748-56.
[12]   Squire LR. Memory and brain systems: 1969-2009. J Neurosci 2009; 29: 12711-6.
[13]   Alberini CM. The role of protein synthesis during the labile phases of memory: revisiting the skepticism. Neurobiol Learn Mem 2008; 89: 234-46.
[14]   Alberini CM. Transcription factors in long-term memory and synaptic plasticity. Physiol Rev 2009; 89: 121-45.
[15]   Gold PE. Protein synthesis and memory. Introduction. Neurobiol Learn Mem 2008; 89: 199-200.
[16]   Silva AJ, Kogan JH, Frankland PW, Kida S. CREB and memory. Annu Rev Neurosci 1998; 21: 127-48.
[17]   Martin KC, Kandel ER. Cell adhesion molecules, CREB, and the formation of new synaptic connections. Neuron 1996; 17: 567-70.
[18]   Johannessen M, Delghandi MP, Moens U. What turns CREB on? Cell Signal 2004; 16: 1211-27.
[19]   Costa-Mattioli M, Sossin WS, Klann E, Sonenberg N. Translational control of long-lasting synaptic plasticity and memory. Neuron 2009; 61: 10-26.
[20]   Agranoff BW, Davis RE, Casola L, Lim R. Actinomycin D blocks formation of memory of shock-avoidance in goldfish. Science 1967; 158: 1600-1.
[21]   Agranoff BW, Davis RE. Further Studies on Memory Formation in the Goldfish. Science 1967; 158: 523.
[22]   Abraham WC, Dragunow M, Tate WP. The role of immediate early genes in the stabilization of long-term potentiation. Mol Neurobiol 1991; 5: 297-314.
[23]   Baeuerle PA, Baltimore D. NF-kappa B: ten years after. Cell 1996; 87: 13-20.
[24]   Baldwin AS, Jr. The NF-kappa B and I kappa B proteins: new discoveries and insights. Annu Rev Immunol 1996; 14: 649-83.
[25]   Hayden MS, Ghosh S. Signaling to NF-kappaB. Genes Dev 2004; 18: 2195-224.
[26]   Kaltschmidt B, Widera D, Kaltschmidt C. Signaling *via* NF-kappaB in the nervous system. Biochim Biophys Acta 2005; 1745: 287-99.
[27]   Lilienbaum A, Israel A. From calcium to NF-kappa B signaling pathways in neurons. Mol Cell Biol 2003; 23: 2680-98.
[28]   Mercurio F, Zhu H, Murray BW, *et al.* IKK-1 and IKK-2: cytokine-activated IkappaB kinases essential for NF-kappaB activation. Science 1997; 278: 860-6.
[29]   Chen LF, Greene WC. Shaping the nuclear action of NF-kappaB. Nat Rev Mol Cell Biol 2004; 5: 392-401.
[30]   Nelsen B, Hellman L, Sen R. The NF-kappa B-binding site mediates phorbol ester-inducible transcription in nonlymphoid cells. Mol Cell Biol 1988; 8: 3526-31.
[31]   Sen R, Baltimore D. Inducibility of kappa immunoglobulin enhancer-binding protein Nf-kappa B by a posttranslational mechanism. Cell 1986; 47: 921-8.
[32]   Kaltschmidt C, Kaltschmidt B, Neumann H, *et al.* Constitutive NF-kappa B activity in neurons. Mol Cell Biol 1994; 14: 3981-92.
[33]   Mattson MP. NF-kappaB in the survival and plasticity of neurons. Neurochem Res 2005; 30: 883-93.
[34]   Massa PT, Aleyasin H, Park DS, *et al.* NFkappaB in neurons? The uncertainty principle in neurobiology. J Neurochem 2006; 97: 607-18.
[35]   Meberg PJ, Kinney WR, Valcourt EG, Routtenberg A. Gene expression of the transcription factor NF-kappa B in hippocampus: regulation by synaptic activity. Brain Res Mol Brain Res 1996; 38: 179-90.

[36]   Freudenthal R, Romano A. Participation of Rel/NF-kappaB transcription factors in long-term memory in the crab Chasmagnathus. Brain Res 2000; 855: 274-81.

[37]   Freudenthal R, Locatelli F, Hermitte G, *et al*. Kappa-B like DNA-binding activity is enhanced after spaced training that induces long-term memory in the crab Chasmagnathus. Neurosci Lett 1998; 242: 143-6.

[38]   Merlo E, Freudenthal R, Romano A. The IkappaB kinase inhibitor sulfasalazine impairs long-term memory in the crab Chasmagnathus. Neuroscience 2002; 112: 161-72.

[39]   Merlo E, Freudenthal R, Maldonado H, Romano A. Activation of the transcription factor NF-kappaB by retrieval is required for long-term memory reconsolidation. Learn Mem 2005; 12: 23-9.

[40]   Albensi BC, Mattson MP. Evidence for the involvement of TNF and NF-kappaB in hippocampal synaptic plasticity. Synapse 2000; 35: 151-9.

[41]   Freudenthal R, Romano A, Routtenberg A. Transcription factor NF-kappaB activation after in vivo perforant path LTP in mouse hippocampus. Hippocampus 2004; 14: 677-83.

[42]   Kassed CA, Willing AE, Garbuzova-Davis S, *et al*. Lack of NF-kappaB p50 exacerbates degeneration of hippocampal neurons after chemical exposure and impairs learning. Exp Neurol 2002; 176: 277-88.

[43]   Freudenthal R BM, Acosta GB, Blake MG, Merlo E, Baratti CM, Romano A. NF-kappaB transcription factor is required for inhibitory avoidance long-term memory in mice. Eur J Neurosci 2005; 24: 1507-16.

[44]   Yeh SH, Lin CH, Lee CF, Gean PW. A requirement of nuclear factor-kappaB activation in fear-potentiated startle. J Biol Chem 2002; 277: 46720-9.

[45]   Yeh SH, Lin CH, Gean PW. Acetylation of nuclear factor-kappaB in rat amygdala improves long-term but not short-term retention of fear memory. Mol Pharmacol 2004; 65: 1286-92.

[46]   Ahn HJ HC, Levenson JM, Lubin FD, Liou HC, Sweatt JD. c-Rel, an NF-kappaB family transcription factor, is required for hippocampal long-term synaptic plasticity and memory formation. Learn Mem 2008; 15: 539-49.

[47]   Meffert MK, Chang JM, Wiltgen BJ, *et al*. NF-kappa B functions in synaptic signaling and behavior. Nat Neurosci 2003; 6: 1072-8.

[48]   O'Mahony A, Raber J, Montano M, *et al*. NF-kappaB/Rel regulates inhibitory and excitatory neuronal function and synaptic plasticity. Mol Cell Biol 2006; 26: 7283-98.

[49]   Piva R, Belardo G, Santoro MG. NF-kappaB: a stress-regulated switch for cell survival. Antioxid Redox Signal 2006; 8: 478-86.

[50]   Pahl HL. Activators and target genes of Rel/NF-kappaB transcription factors. Oncogene 1999; 18: 6853-66.

[51]   Taylor CT, Cummins EP. The role of NF-kappaB in hypoxia-induced gene expression. Ann N Y Acad Sci 2009; 1177: 178-84.

[52]   Albensi BC. Potential roles for tumor necrosis factor and nuclear factor-kappaB in seizure activity. J Neurosci Res 2001; 66: 151-4.

[53]   Unlap T, Jope RS. Inhibition of NFkB DNA binding activity by glucocorticoids in rat brain. Neurosci Lett 1995; 198: 41-4.

[54]   Malek R, Borowicz KK, Jargiello M, Czuczwar SJ. Role of nuclear factor kappaB in the central nervous system. Pharmacol Rep 2007; 59: 25-33.

[55]   Furukawa K, Mattson MP. The transcription factor NF-kappaB mediates increases in calcium currents and decreases in NMDA- and AMPA/kainate-induced currents induced by tumor necrosis factor-alpha in hippocampal neurons. J Neurochem 1998; 70: 1876-86.

[56]   Marini AM, Jiang X, Wu X, *et al*. Role of brain-derived neurotrophic factor and NF-kappaB in neuronal plasticity and survival: From genes to phenotype. Restor Neurol Neurosci 2004; 22: 121-30.

[57]   Ohba S, Ikeda T, Ikegaya Y, *et al*. BDNF locally potentiates GABAergic presynaptic machineries: target-selective circuit inhibition. Cereb Cortex 2005; 15: 291-8.

[58]   Yu Z, Cheng G, Wen X, *et al*. Tumor necrosis factor alpha increases neuronal vulnerability to excitotoxic necrosis by inducing expression of the AMPA-glutamate receptor subunit GluR1 *via* an acid sphingomyelinase- and NF-kappaB-dependent mechanism. Neurobiol Dis 2002; 11: 199-213.

[59]   O'Neill LA, Kaltschmidt C. NF-kappa B: a crucial transcription factor for glial and neuronal cell function. Trends Neurosci 1997; 20: 252-8.

[60]   McGaugh JL. Memory--a century of consolidation. Science 2000; 287: 248-51.

[61]   Castellucci VF, Blumenfeld H, Goelet P, Kandel ER. Inhibitor of protein synthesis blocks long-term behavioral sensitization in the isolated gill-withdrawal reflex of Aplysia. J Neurobiol 1989; 20: 1-9.

[62]   Davis HP, Squire LR. Protein synthesis and memory: a review. Psychol Bull 1984; 96: 518-59.

[63]   Bliss TV, Collingridge GL. A synaptic model of memory: long-term potentiation in the hippocampus. Nature 1993; 361: 31-9.

[64]   Cole AJ, Saffen DW, Baraban JM, Worley PF. Rapid increase of an immediate early gene messenger RNA in hippocampal neurons by synaptic NMDA receptor activation. Nature 1989; 340: 474-6.

[65]   Wisden W, Errington ML, Williams S, *et al*. Differential expression of immediate early genes in the hippocampus and spinal cord. Neuron 1990; 4: 603-14.

[66]   Baumeister MA, Rossman KL, Sondek J, Lemmon MA. The Dbs PH domain contributes independently to membrane targeting and regulation of guanine nucleotide-exchange activity. Biochem J 2006; 400: 563-72.

[67]   Paylor R, Johnson RS, Papaioannou V, *et al*. Behavioral assessment of c-fos mutant mice. Brain Res 1994; 651: 275-82.

[68]   Plath N, Ohana O, Dammermann B, *et al*. Arc/Arg3.1 is essential for the consolidation of synaptic plasticity and memories. Neuron 2006; 52: 437-44.

[69]   Ploski JE, Pierre VJ, Smucny J, *et al*. The activity-regulated cytoskeletal-associated protein (Arc/Arg3.1) is required for memory consolidation of pavlovian fear conditioning in the lateral amygdala. J Neurosci 2008; 28: 12383-95.

[70]   Frey U, Huang YY, Kandel ER. Effects of cAMP simulate a late stage of LTP in hippocampal CA1 neurons. Science 1993; 260: 1661-4.

[71]   Huang YY, Kandel ER. Recruitment of long-lasting and protein kinase A-dependent long-term potentiation in the CA1 region of hippocampus requires repeated tetanization. Learn Mem 1994; 1: 74-82.

[72]   Huang YY, Li XC, Kandel ER. cAMP contributes to mossy fiber LTP by initiating both a covalently mediated early phase and macromolecular synthesis-dependent late phase. Cell 1994; 79: 69-79.

[73]   Matthies H, Reymann KG. Protein kinase A inhibitors prevent the maintenance of hippocampal long-term potentiation. Neuroreport 1993; 4: 712-4.

[74]   Nguyen PV, Abel T, Kandel ER. Requirement of a critical period of transcription for induction of a late phase of LTP. Science 1994; 265: 1104-7.

[75]   Ghirardi M, Braha O, Hochner B, *et al*. Roles of PKA and PKC in facilitation of evoked and spontaneous transmitter release at depressed and nondepressed synapses in Aplysia sensory neurons. Neuron 1992; 9: 479-89.

[76]   Dash PK, Hochner B, Kandel ER. Injection of the cAMP-responsive element into the nucleus of Aplysia sensory neurons blocks long-term facilitation. Nature 1990; 345: 718-21.

[77]   Alberini CM, Ghirardi M, Metz R, Kandel ER. C/EBP is an immediate-early gene required for the consolidation of long-term facilitation in Aplysia. Cell 1994; 76: 1099-114.

[78]   Glanzman DL, Kandel ER, Schacher S. Target-dependent structural changes accompanying long-term synaptic facilitation in Aplysia neurons. Science 1990; 249: 799-802.

[79]   Nazif FA, Byrne JH, Cleary LJ. cAMP induces long-term morphological changes in sensory neurons of Aplysia. Brain Res 1991; 539: 324-7.

[80]   Kaltschmidt B, Ndiaye D, Korte M, *et al.* NF-kappaB regulates spatial memory formation and synaptic plasticity through protein kinase A/CREB signaling. Mol Cell Biol 2006; 26: 2936-46.

[81]   Milner B, Squire LR, Kandel ER. Cognitive neuroscience and the study of memory. Neuron 1998; 20: 445-68.

[82]   Brightwell JJ, Smith CA, Neve RL, Colombo PJ. Transfection of mutant CREB in the striatum, but not the hippocampus, impairs long-term memory for response learning. Neurobiol Learn Mem 2008; 89: 27-35.

[83]   Herdegen T, Leah JD. Inducible and constitutive transcription factors in the mammalian nervous system: control of gene expression by Jun, Fos and Krox, and CREB/ATF proteins. Brain Res Brain Res Rev 1998; 28: 370-490.

[84]   Kandel ER. Genes, synapses, and long-term memory. J Cell Physiol 1997; 173: 124-5.

[85]   Impey S, Mark M, Villacres EC, *et al.* Induction of CRE-mediated gene expression by stimuli that generate long-lasting LTP in area CA1 of the hippocampus. Neuron 1996; 16: 973-82.

[86]   Deisseroth K, Bito H, Tsien RW. Signaling from synapse to nucleus: postsynaptic CREB phosphorylation during multiple forms of hippocampal synaptic plasticity. Neuron 1996; 16: 89-101.

[87]   Abel T, Nguyen PV, Barad M, *et al.* Genetic demonstration of a role for PKA in the late phase of LTP and in hippocampus-based long-term memory. Cell 1997; 88: 615-26.

[88]   Bading H, Ginty DD, Greenberg ME. Regulation of gene expression in hippocampal neurons by distinct calcium signaling pathways. Science 1993; 260: 181-6.

[89]   Barrett RM, Wood MA. Beyond transcription factors: the role of chromatin modifying enzymes in regulating transcription required for memory. Learn Mem 2008; 15: 460-7.

[90]   Routtenberg A, Rekart JL. Post-translational protein modification as the substrate for long-lasting memory. Trends Neurosci 2005; 28: 12-9.

[91]   Choi DW. Excitotoxic cell death. J Neurobiol 1992; 23: 1261-76.

[92]   Bojarski L, Herms J, Kuznicki J. Calcium dysregulation in Alzheimer's disease. Neurochem Int 2008; 52: 621-33.

[93]   Canzoniero LM, Snider BJ. Calcium in Alzheimer's disease pathogenesis: too much, too little or in the wrong place? J Alzheimers Dis 2005; 8: 147-54; discussion 209-15.

[94]   Drouet B, Pincon-Raymond M, Chambaz J, Pillot T. Molecular basis of Alzheimer's disease. Cell Mol Life Sci 2000; 57: 705-15.

[95]   Parsons CG, Stoffler A, Danysz W. Memantine: a NMDA receptor antagonist that improves memory by restoration of homeostasis in the glutamatergic system--too little activation is bad, too much is even worse. Neuropharmacology 2007; 53: 699-723.

[96]   Freudenthal R, Boccia MM, Acosta GB, *et al.* NF-kappaB transcription factor is required for inhibitory avoidance long-term memory in mice. Eur J Neurosci 2005; 21: 2845-52.

[97]   Dash PK, Orsi SA, Moore AN. Sequestration of serum response factor in the hippocampus impairs long-term spatial memory. J Neurochem 2005; 93: 269-78.

[98]   Collister KA, Albensi BC. Potential therapeutic targets in the NF-kappaB pathway for Alzheimer's disease. Drug News Perspect 2005; 18: 623-9.

[99]   Sarnico I, Lanzillotta A, Benarese M, *et al.* NF-kappaB dimers in the regulation of neuronal survival. Int Rev Neurobiol 2009; 85: 351-62.

[100]   Romano A, Freudenthal R, Merlo E, Routtenberg A. Evolutionarily-conserved role of the NF-kappaB transcription factor in neural plasticity and memory. Eur J Neurosci 2006; 24: 1507-16.

[101]   Bhakar AL, Tannis LL, Zeindler C, *et al.* Constitutive nuclear factor-kappa B activity is required for central neuron survival. J Neurosci 2002; 22: 8466-75.

[102]   Koulich E, Nguyen T, Johnson K, *et al.* NF-kappaB is involved in the survival of cerebellar granule neurons: association of IkappaBbeta [correction of Ikappabeta] phosphorylation with cell survival. J Neurochem 2001; 76: 1188-98.

[103]   Maggirwar SB, Sarmiere PD, Dewhurst S, Freeman RS. Nerve growth factor-dependent activation of NF-kappaB contributes to survival of sympathetic neurons. J Neurosci 1998; 18: 10356-65.

[104]   Middleton G, Hamanoue M, Enokido Y, *et al.* Cytokine-induced nuclear factor kappa B activation promotes the survival of developing neurons. J Cell Biol 2000; 148: 325-32.

[105]   Lisman JE, Harris KM. Quantal analysis and synaptic anatomy--integrating two views of hippocampal plasticity. Trends Neurosci 1993; 16: 141-7.

[106]   Gutierrez H, Hale VA, Dolcet X, Davies A. NF-kappaB signalling regulates the growth of neural processes in the developing PNS and CNS. Development 2005; 132: 1713-26.

[107]   Denis-Donini S, Dellarole A, Crociara P, *et al.* Impaired adult neurogenesis associated with short-term memory defects in NF-kappaB p50-deficient mice. J Neurosci 2008; 28: 3911-9.

[108]   Perera TD, Coplan JD, Lisanby SH, *et al.* Antidepressant-induced neurogenesis in the hippocampus of adult nonhuman primates. J Neurosci 2007; 27: 4894-901.

[109]   Santarelli L, Saxe M, Gross C, *et al.* Requirement of hippocampal neurogenesis for the behavioral effects of antidepressants. Science 2003; 301: 805-9.

[110]   Karin M, Yamamoto Y, Wang QM. The IKK NF-kappa B system: a treasure trove for drug development. Nat Rev Drug Discov 2004; 3: 17-26.

[111]   Yin MJ, Yamamoto Y, Gaynor RB. The anti-inflammatory agents aspirin and salicylate inhibit the activity of I(kappa)B kinase-beta. Nature 1998; 396: 77-80.

[112]   Keifer JA, Guttridge DC, Ashburner BP, Baldwin AS, Jr. Inhibition of NF-kappa B activity by thalidomide through suppression of IkappaB kinase activity. J Biol Chem 2001; 276: 22382-7.

[113]   Poirier R, Cheval H, Mailhes C, *et al.* Distinct functions of egr gene family members in cognitive processes. Front Neurosci 2008; 2: 47-55.

[114]   Beckmann AM, Wilce PA. Egr transcription factors in the nervous system. Neurochem Int 1997; 31: 477-510; discussion 517-6.

CHAPTER 5

NF-κB Proteins in Adult Neurogenesis: Relevance for Learning and Memory in Physiology and Pathology

Mariagrazia Grilli[1,2*] and Vasco Meneghini[1,2]

[1]*DiSCAFF, University of Piemonte Orientale "A. Avogadro", and* [2]*DFB Center, Laboratory of Neuroplasticity & Pain, Novara 28100, Italy*

Abstract: An increasing number of reports have called attention on the role of the NF-κB family members in processes that require long-term regulation of the synaptic function and underlying memory and neural plasticity. In general, NF-κB proteins appear to participate in spatial and contextual memory tasks that depend on the hippocampal function, whereas nonspatial or cued tasks that are not dependent on hippocampus usually show little impairment in NF-κB deficient animals.

The molecular mechanisms and target genes through which the complex NF-κB pathway mediates transcriptional regulation to participate in memory and learning are far less characterized. Recently, a potential involvement of NF-κB proteins in adult neurogenesis, the formation of new neurons in adulthood, has been suggested. While its exact function remains elusive, adult hippocampal neurogenesis has been proposed to play important roles in specific cognitive functions and to contribute to their maintenance. In particular it appears that neurogenesis may be relevant for some hippocampal-dependent memory tasks. Whether the two phenomena, NF-κB-mediated regulation of cognitive function and modulation of adult neurogenesis are functionally linked remains to be unambiguously confirmed. Nevertheless these findings are herein reviewed and discussed for their impact in several neuropsychiatric disorders and their therapeutic implications.

Keywords: NF-κB, neurogenesis, hippocampus, neural stem cells, neural progenitor cells, subgranular zone, learning, memory, neuroplasticity, neurodegeneration, neuropsychiatric disorders.

THE NF-κB FAMILY OF TRANSCRIPTION FACTORS

NF-κB was first described in 1986 as a nuclear factor, restricted to B lymphocytes, that specifically bound to a 10 bp sequence located in the enhancer region of the κ immunoglobulin light chain gene and activated its transcription [1]. After two decades, the chosen name (<u>N</u>uclear <u>F</u>actor in the <u>kappa</u> light chain enhancer of <u>B</u> cells) appears a misnomer since NF-κB proteins are now recognized as ubiquitously expressed transcription factors that control expression of a large variety of genes in any given tissue. The ever growing list of target genes, although mainly on non-neuronal cells, is updated on T.D. Gilmore's webpage (www.nf-kb.org). Since their discovery, NF-κB proteins have been attributed a plethora of different functions, their role ranging from immunity and host defence to apoptosis and cell survival, cellular growth and repair processes, oncogenesis, embryonic patterning [2-3]. Part of the NF-κB family pleiotropicity relies on the molecular complexity within the system. Five mammalian NF-κB DNA-binding subunits that can form homo or heterodimeric combinations have been identified: p50 (NF-κB1), p52 (NF-κB2), p65 (Rel-A), c-Rel, and Rel B, all sharing the Rel homology domain (RHD) [4]. The first two are also synthesized as large precursors of 105 kDa (p105) and 100 kDa (p100), respectively, and are partially proteolyzed by the 26S proteasome, resulting in mature units. The regulatory specificity of hetero- and homo-dimers within the family is also dependent on their distinct affinities for κB consensus sequences as well as for cofactors and this

Address correspondence to Mariagrazia Grilli: Laboratory of Neuroplasticity and Pain, DiSCAFF & DFB Center, Via Bovio 6, 28100 Novara, Italy; Tel: +390321375828; Fax: +390321375821; E-mail: grilli@pharm.unipmn.it

in part contributes to the fact that dimers with different composition may elicit multiple, even opposite, functions within the same cell type [5-6]. Additionally, these transcription factors have posttranslationally regulated activity. Inhibitors of NF-κB, called IκB proteins, are able to sequester the NF-κB complex in the cytoplasm. The IκB family includes at least seven members, namely IκBα, IκBβ, IκBε, p105 (and its C-terminally derived IκBγ), p100 (and its C-terminally derived IκBδ), IκBζ, and bcl3, some of them displaying preferences of complex formation with distinct dimers. Nuclear roles for IκB proteins have also been demonstrated, adding another layer of complexity to the regulatory mechanisms that act to activate or repress a wide range of gene targets [7]. In the so-called "canonical pathway", inactivation of IκBα is mediated by its degradation within the cytoplasm. In response to diverse sets of stimuli (such as cytokines and Lipopolysaccaride, LPS), IκB is phosphorylated by the IκB kinase (IKK) complex, consisting of two catalytic subunits IKK1/α, IKK2/β and a regulatory subunit NEMO/IKKγ. Upon stimulation, the IKK complex triggers phosphorylation of two N-terminal serines within the IκBs, leading to their ubiquitination and degradation through the 26S proteasome. In the "alternative pathway", the NF-κB inducing kinase (NIK) activates the homomeric IKKα complex that in turn phosphorylates and partially degrades the NF-κB2 p100-RelB complex into active NF-κB2 p52-RelB. Initially believed to operate only in the immune system upon stimulation by lymphotoxin β, B-cell activating factor (BAFF) or CD40, the alternative pathway is utilized also by astrocytes in the CNS to modulate brain inflammation in experimental autoimmune encephalomyelitis [8]. Even if degradation of IκBs represents a crucial step in NF-κB activation, other regulatory mechanisms, including phosphorylation, acetylation, sumoylation of NF-κB and/or IKK subunits confer additional fine tuning within the system [9-11].

Two subfamilies of NF-κB proteins can be identified: a Rel subfamily comprising RelA, c-Rel and RelB and a NF-κB subfamily including NF-κB1 and NF-κB2. RelA, RelB and c-rel are generally thought to activate transcription because they do contain transactivator domains (TAD), while p50 and p52 homodimers lacking TAD are generally considered as repressors. Again, this is an oversimplified picture since it has been shown that bcl3, and possibly IκBζ, can direct transcriptional activation by forming a ternary complex with DNA through p50 and p52 homodimers [12-13].

Although the NF-κB family has been studied extensively in the context of immunity, inflammation, cancer, far less is understood on the function and regulation of the NF-κB family in the nervous system. Of particular relevance for their therapeuthic implications, a vast array of experimental data underline the importance of the NF-κB system in cognitive functions such as learning and memory and, more recently, also in adult neurogenesis. Whether the two phenomena are functionally linked remains to be unambiguously confirmed but these interesting findings are herein reviewed and discussed for their impact in several neuropsychiatric disorders.

NF-κB ROLE IN SYNAPTIC PLASTICITY IN LEARNING AND MEMORY PROCESSES

In the past, an increasing number of reports have called attention on the role of the NF-κB family members in processes that require long-term regulation of the synaptic function and underlying memory and neural plasticity. Several studies showed that latent NF-κB dimers are present in synaptic regions, and that once activated, NF-κB translocates from the synapse to the nucleus [14-19]. This strongly suggested that the NF-κB transcription factor family may act as a synapse-to-nucleus messenger in the nervous system, in addition to its role as a transcriptional regulator. Moreover these data further substantiated the idea that recruitment of NF-κB from synapse to nucleus could be part of molecular mechanisms involved in memory formation processes. NF-κB subcellular distribution, DNA binding activity, and transcription are indeed regulated by various forms of synaptic activity [16, 18, 20-25]. Moreover pre-exposure to NF-κB decoy oligonucleotides blocks induction of long term potentiation (LTP) in hippocampus [21] and amygdala [26].

Currently the NF-κB contribution to neural plasticity and memory is widely recognized [27-29] and this role appears to be evolutionary-conserved [30]. The use of the gene-targeting strategy in mice has certainly contributed to shed light on the role of NF-κB in learning and memory in mammals. Deletion of the p65 subunit is lethal during embryogenesis [31] but the knock-out can be rescued by concomitant deletion of type 1 TNF receptor (TNFR-1) [32]. The brains of p65/TNFR1 double knock-out mice appear normal but formation of hippocampal-dependent spatial memory (and not of striatal-dependent working memory) is blocked [18]. Moreover, c-Rel knock-out inhibits formation of hippocampus-dependent long-term memory but does not affect amygdala-dependent long-term memory formation [33-34]. Another interesting model is the tTA/IκBα-AA double transgenic mouse in which the inducible Tet-off system under the control of a CamKIIa promoter allows to regulate postnatal expression of a dominant negative IκBα transgene (IκBα-dn) in forebrain regions [35]. NF-κB forebrain-deficient mice were shown to have impaired late LTP and long term depression (LTD) and disrupted formation of spatial memory in the Morris water maze, although no differences were found between NF-κB-deficient animals and doxycycline-treated animals. Additionally, NF-κB-deficient animals performed normally in another non-spatial task, the short-term object recognition memory, which is hippocampus-independent.

Astrocytes play a pivotal role in regulating synaptic plasticity and synapse formation and several stimuli are known to trigger activation of NF-κB in those cells [36]. Bracchi-Ricard [37] investigated the role of astroglial NF-κB in synaptic plasticity and cognition using transgenic mice expressing IκBα-dn under control of the glial fibrillary acidic protein (GFAP) that generates expression in astrocytes as well as progenitor cells. GFAP-IκBα-dn male and female mice and their corresponding wild-type (wt) littermates were analysed for two different learning tasks, the non-cued Barnes maze and the contextual fear conditioning paradigm. Interestingly, female but not male mice, were significantly impaired in both spatial learning tasks, suggesting a sexual dimorphism in NF-κB signaling in astrocytes or progenitor cells. Additionally, female transgenic mice were also impaired in a non-spatial learning test pointing to an extra-hippocampal component. The behavioral deficits in female transgenic mice were also accompanied by a hippocampal deficit in LTP and reduced expression of metabotropic glutamate receptor mGluR5 in hippocampus and cortex, and postsynaptic density protein 95 (PSD95) in the cortex [37].

Mice with a p50 null mutation are viable and fertile and do not have gross developmental and histological defects, although they show immune deficits [38]. The mutant mice also displayed acquisition deficits in an active avoidance task but no differences were found between p50$^{-/-}$ and wt mice in a test for long-term memory, suggesting that p50 is not required for memory formation [39]. Another study characterized these mice at the behavioural level, showing decreased anxiety-like responses, elevated exploratory behaviour and a reduced tendency to develop dominant-subordinate behaviours [40]. Such relevant behavioural differences between mutants and wt animals could possibly account for the differences in learning performance in active avoidance.

Recent studies have identified novel roles for signaling components of the NF-κB pathway, such as IκBα and IKKα, in regulating chromatin structure and controlling gene expression independent of NF-κB binding to regulatory DNA elements [41-43]. Although these regulatory mechanisms have been mainly investigated in non neuronal cells, some studies have specifically suggested that epigenetic mechanisms, such as chromatin remodelling through histone regulation, may be critical for normal synaptic plasticity and for triggering long-term changes in neuronal function [44-45]. Lubin & Sweatt [46] reported that post-translational modifications (phosphorylation and acetylation) of the histone H3-tail occur in the hippocampus after recall of contextual conditioned fear memory and that this regulation required activation of the NF-κB pathway *via* IKKα and not the NF-κB DNA-binding complex. Overall, these findings demonstrate that activation of the IKK-regulated transcription mechanism at the level of chromatin structure in hippocampus may also be important for memory reconsolidation in the mammalian CNS.

In conclusion, a vast array of information has been collected to suggest an important role for NF-κB in activity-dependent synaptic plasticity of the hippocampal formation. In general, NF-κB proteins appear to participate in spatial and contextual memory tasks that depend on the hippocampal function, whereas nonspatial or cued tasks that are not dependent on hippocampus usually show little impairment in NF-κB deficient animals. Overall these studies suggest that specific mechanisms exist for activation of the NF-κB transcriptional pathway during various stages of memory formation. However, the molecular mechanisms and target genes through which the complex NF-κB pathway mediates transcriptional regulation to stabilize memory are far less characterized. Unfortunately, available studies aimed at identifying NF-κB target genes have been performed mainly in cell lines and often those lines are not even CNS-derived. Only a few studies have attempted to study NF-κB target genes in the context of memory formation [33,35] and many more are needed in the future.

ADULT NEUROGENESIS

Neurogenesis, or the birth of new neuronal cells, was originally thought to occur only in developing organisms. More recently, a vast array of experimental work has firmly established that new neurons are generated throughout life from Neural Stem Cells (NSC) in discrete regions of the adult brain of mammals, including humans [47-50]. NSC are cells endowed with long-term self-renewal capacity and ability to differentiate into multiple cell phenotypes, including neurons. The active neurogenic regions in adulthood are the Sub Granular Zone (SGZ) of the dentate gyrus (DG), which generates hippocampal neurons, and the SubVentricular Zone (SVZ) lining the walls of the lateral ventricles, which generates interneurons that migrate into the olfactory bulb (OB). In particular, within the hippocampus, new dentate granule cells are continuously generated from Neural Progenitor Cells (NPC) and integrated into the existing circuitry through an orchestrated process. The events that contribute to hippocampal postnatal neurogenesis rely on the maintenance and proliferation of multipotent stem/progenitor cells located in the SGZ niche, a permissive and instructive microenvironment where NSC survive throughout life, self-renew and give rise, by asymmetric division, to precursors cells which then migrate into the adjacent granule cell layer (GCL). Only a small proportion of these committed precursors survive and differentiate into neurons, while the majority die [51]. Adult generated newborn neurons extend axons and dendrites and are then integrated into the preexisting functional network [48, 52]. In summary, net adult neurogenesis results from the complex interplay of cell proliferation and cell death, migration and differentiation and the molecular pathways that underlie these cellular processes are obviously multifaceted. Additionally, close interactions of putative stem cells with transient precursor cells have been observed in neurogenic regions, suggesting that primary progenitors continue to interact closely with their progeny, possibly exchanging signals that are important for their differentiation, maturation and survival [47-49]. Since in CNS NF-κB proteins have been shown to be involved in functions such as cell proliferation, differentiation, migration, survival, neuron-glia communication [18, 22, 27-29], they could also represent potential candidates for translating complex signals which regulate the rate of neurogenesis in the highly specialized neurogenic niches.

LINKS BETWEEN LEARNING/MEMORY AND ADULT HIPPOCAMPAL NEUROGENESIS

Despite great excitement regarding the phenomenon, the functional contribution of adult neurogenesis to the neuronal network remains largely unknown and extensively debated. Data, primarily from correlational or ablation studies have given rise to controversial and sometimes inconsistent findings across laboratories and animal species. Altogether, several groups have reported data correlating reduced/enhanced neurogenesis with deficit/enhancement in learning and memory performance [53-61]. While the exact function remains elusive, adult hippocampal neurogenesis has been suggested to play important roles in specific cognitive functions and to contribute to their maintenance. In particular it appears that neurogenesis is important for at least some hippocampal-dependent memory tasks. Recent elegant studies on the functional role of these cells have used immediate-early gene expression as a readout of adult-generated neuron activation [59, 62] or analysis of stabilization and removal of new neurons of different ages after a hippocampal-dependent learning task [63]. The preferential incorporation of adult-generated granule cells into spatial memory networks has been suggested [59]. Recently, Dupret *et al.* [63] have proposed that spatial learning requires both addition and removal of the newly generated neurons, providing evidence of an even more complex mechanism contributing to memory formation in mammals. However, much remains unknown about the responses of new neurons to hippocampal-dependent learning tasks and their contribution to hippocampal function. To determine how learning activates newly generated granule cells, Snyder *et al.* [64] trained adult rats in a hippocampal-dependent task such as the Morris water maze, and evaluated expression of immediate-early genes in adult-generated hippocampal neurons. They could show that adult-generated neurons are activated during learning, and that this activation depends on the amount or strength of training as well as the variations in events during training. They also found that hippocampally-relevant experience activates young and mature neurons in different DG subregions and with different experiential specificity. A remarkable feature of adult hippocampal neurogenesis is that it is regulated by a variety of conditions. A range of environmental stimuli have profound effects on proliferation and differentiation of neural progenitors as well as survival of their progeny in the adult DG. Both environmental enrichment (EE) and voluntary physical exercise promote hippocampal neurogenesis and EE/exercise-induced increase in adult hippocampal neurogenesis was correlated with enhanced long-term potentiation in the dentate gyrus and improved spatial learning [55, 65-68]. On the other hand, reduction in hippocampal neurogenesis due to aging and stress is hypothesized to contribute to the cognitive decline observed in these settings [69-74].

DISREGULATED NF-κB SIGNALING AND NEUROGENESIS IN CNS DISORDERS

Neurogenesis in the adult brain is altered in various CNS disorders, including neurodegenerative diseases, stroke, trauma, epilepsy, where the NF-κB signaling pathway is also disregulated.

One post-mortem study of senile Alzheimer's Disease (AD) brains demonstrated an increase in the number of cells expressing markers for cell proliferation and of immature neurons [75], possibly suggesting an up-regulation of neurogenesis. Conversely, another study of the presenile AD brain revealed that most of the increased number of proliferating cells in patients are non-neuronal [76]. In animal models of AD, data are also contradictory. Discrepancies may originate from the limitations of animal models as representative models of complex diseases, such as AD and/or by distinct effects of Familial AD-associated genes such as Amyloid Precursor Protein (APP) and Presenilins on neurogenesis in double or triple transgenic mouse lines. *In vitro* beta amyloid (Aβ) is reported to affect cultured neural stem/progenitor cells by suppressing their proliferation and neuronal differentiation and by promoting apoptosis [77]. The effect of Aβ on neurogenesis has been shown in mice given Aβ intraventricularly. In these mice, neuronal production in the DG by neural stem/progenitor cells and the survival/differentiation of the newly generated neurons are impaired [78]. Although there is no cure for AD, cholinesterase inhibitors (ChEI) can slow down or transiently improve the cognitive dysfunction in patients. ChEI, which ameliorate the depletion of acetylcholine (ACh) by inhibiting the degradation of synaptic ACh by cholinesterase, were originally developed with the intent of supporting cholinergic neuronal transmission. However, recent data suggest that some of these drugs may have additional, neuroprotective effects. Interestingly, long-term administration of the ChEI donepezil was shown to promote the survival of newly generated neurons and to increase neurogenesis in both the DG and the OB [79]. Whether these effects may contribute to the functional improvement in memory and cognition seen in AD with the use of this drug it remains to be established. Interestingly, AD mutant mice living in an enriched environment, which promotes adult hippocampal neurogenesis, also show significant improvement in memory and cognitive tasks compared with mice bearing the same mutation and living in standard cages [80].

Controversial data are also available on the alterations of the NF-κB signaling pathway in AD. Studies in postmortem brain tissues from AD patients revealed augmented p65 immunoreactivity in cholinergic neurons of the basal forebrain as well as in hippocampus and enthorinal cortex [81-83]. Conversely, a strong decrease of p65 immunoreactivity was reported in cells surrounding amyloid plaques of AD patients in comparison to healthy controls [84]. These controversial data were interpreted as if NF-κB activation may exert both neuroprotective and neurotoxic roles in AD. *In vitro* data are also of complex interpretation. In neuroblastoma and in primary rat cerebellar granule cells, Aβ peptide triggers NF-κB nuclear translocation [81, 85-86]. At high concentrations, the amyloidogenic peptide promoted neuronal apoptosis mediated by p50/p65 nuclear translocation [87]. Conversely, pre-activation of NF-κB by tumor necrosis factor (TNF) or low concentrations of Aβ protected neurons against the cytotoxic effects induced by high concentrations of amyloidogenic peptides [84, 88]. Moreover, the anti-apoptotic effects elicited by mGluR5 receptor agonists against neuronal Aβ toxicity were mediated by activation of c-Rel containing dimers which upregulated Bcl-xL and manganese superoxide dismutase (MnSOD) levels [89]. NF-κB activation following exposure to sAPPα counteracted the pro-apoptotic effects of mutated presenilin-1 (PS1) [90]. Interestingly, an aberrant pattern of NF-κB activation was observed in cells expressing mutant PS1 after exposure to oxidative insults, consisting in an enhanced early activation followed by a prolonged inhibition, which could be restored by pretreating cells with sAPPα [90]. Overall literature data suggest that the role of NF-κB in neuronal survival or death might be dependent on several factors including the timing and the subunit composition of dimeric complexes which are activated. In particular, when Aβ is the inducing stimulus, an early activation of NF-κB dimers containing c-Rel seems to exert neuroprotective effects, whereas a later involvement of NF-κB p50/p65 dimers seems to induce Aβ neurotoxicity. Accordingly, only the p50/p65 dimer was activated in the first hours of exposure to neurotoxic Aβ concentrations [87], whereas silencing of c-Rel or its absence in c-Rel$^{-/-}$ neurons abolished the mGluR5-mediated protection against the amyloidogenic peptide [89]. NF-κB-mediated transcription plays also an important role in neuroinflammation, a component of AD pathogenesis. As an example, in neurons, the activation of the receptor for advanced glycation end-products (RAGE) by Aβ peptides elicited synthesis and secretion of macrophage-colony stimulating factor (M-CSF) which in turn promoted microglia activation [86]. Moreover, low concentrations of Aβ stimulated NF-κB-mediated production of inflammatory mediators (IL-6, IL-8 and NOS-II) in post-mortem human cortical microglia [91]. Interestingly, the expression of IκBα-dn in microglia inhibited Aβ neurotoxicity [92] suggesting a potential role of microglial NF-κB signaling in peptide neurotoxicity.

In Hungtington's disease (HD) and Parkinson's disease (PD) main alterations in neurogenesis are in the adult SVZ region. Post-mortem analyses of brains from HD patients have shown that the SVZs have become thicker, with increased cell proliferation, mainly of the neural stem cells [93-94]. These alterations are reproduced in an HD transgenic mouse model, R6/2 mice, which carries human HD genes with long CAG repeats. The ability of neural stem cells dissociated from the SVZ of these mice to self-renew increases in parallel with the progression of the disease [95]. The administration of basic Fibroblast Growth factor (bFGF), a major trophic factor that up-regulates SVZ neurogenesis, to R6/2 mice causes an increase in SVZ cell proliferation of about 150%, as well as increased migration of neuroblasts to the striatum and the regeneration of striatal projection neurons [96]. Moreover, this treatment ameliorates the motor dysfunction and extends the life-span of these mice [96]. These findings suggest that the altered neurogenesis that occurs during HD could represent a protective response of the central nervous system, and that interventions that enhance SVZ neurogenesis could be beneficial to this progressive disease. Additionally, some studies report that, unlike in the SVZ, cell proliferation and neurogenesis in the hippocampus of HD mouse models R6/1 and R6/2 mice is decreased [97-98]. The relationship between the decreased hippocampal neurogenesis and the neuropathology and cognitive deficits associated with HD has not been defined. However, it is interesting that also in this disease model environmental enrichment and exercise delayed the progression of HD symptoms [99].

Several lines of evidence suggest that NF-κB may regulate polyglutamine-induced neurodegeneration in HD. In HD transgenic mice, nuclear NF-κB was detected in cortical and striatal neurons and its inhibition significantly reduced mutant hungtingtin (Htt)-induced neurotoxicity in cell cultures and in brain slices [100]. In addition, mutant Htt activated the IKK complex through direct association with NEMO in PC12 cells [100]. NF-κB activation and expression of pro-apoptotic genes, such as p53, were observed in medium-size striatal neurons of mice injected with the mitochondrial toxin 3-nitropropionic acid (3PN), which mimics the neurodegeneration induced by mutant Htt [101]. Altogether these evidences would indicate that NF-κB may play a promoting role in neurodegeneration associated with HD. In contrast, p50-deficient mice injected with 3PN display increased loss of striatal neurons [102], suggesting a more complex role of NF-κB signaling in HD.

In contrast to the increased SVZ proliferation seen in HD, cell proliferation in the SVZ of PD patients and animal models is significantly suppressed. Höglinger and colleagues [103] showed that the proliferation of transit-amplifying cells in the SVZ and of neuronal progenitors in the DG is positively controlled by the dopaminergic input of the nigrostriatal afferents, which are the cells that are selectively damaged in PD. Interestingly, the decrease in the numbers of neural stem cells and neuroblasts in the SVZ is more significant in PD patients with cognitive impairments (PD with dementia) than in those without [103]. Thus, decreased neurogenesis in PD brain are potentially attributable to the loss of dopaminergic neurons, and may underlie additional symptoms, including cognitive disturbances. A vast array of data suggested alterations of the NF-κB signaling in the human disorder. Indeed a 70-fold increase in the proportion of dopaminergic neurons exhibiting nuclear p65 immunoreactivity was observed in the substantia nigra of PD patients compared to age-matched controls [104]. *In vitro* studies have proposed that production of free radicals, which both activates NF-κB and triggers neuronal death of dopaminergic neurons, might be one the mechanisms underlying neurodegeneration in PD [105-108]. But the role of NF-κB in dopaminergic neurons is much more controversial. NF-κB activation has been shown to protect the dopaminergic neuronal MN9D cells by 6-OHDA-induced cell death [109]. Moreover, based on results obtained in $p50^{-/-}$ mice, it was proposed that NF-κB signaling may play a minor role in the 1-methyl-4-phenyl-1,2,3,6-tetrahydropteridine (MPTP) mouse model of PD [110].

A recent study on post-mortem time revealed that stroke patients produced new neurons following the insult [111]. Moreover, Liu *et al.* [112] reported that neurogenesis in the DG is markedly increased in an adult gerbil model of transient global ischemia, which causes the death of pyramidal neurons in the CA1 region of the hippocampus; however, CA1 neurons were not replaced. In contrast, another group [113] demonstrated the regeneration of CA1 pyramidal neurons in an adult rat model for transient global ischemia. In rodents, middle cerebral artery occlusion

(MCAO) is the most common model for ischemic stroke that causes infarction in the striatum and adjacent parietal cortex. Within a week after the insult, neural stem/progenitor cells in the SVZ begin to proliferate, and neuroblasts with the migratory morphology and newly generated neurons appear at the boundary of the damaged area in the striatum [114]. By using viral infection-mediated cell-specific introduction of GFAP expression, Yamashita *et al.* [115] showed that these neurons are generated by GFAP expressing neural stem cells in the SVZ and that they migrate radially into the damaged striatum, where they differentiate into mature neurons. Also in stroke, these data would suggest that the continuous production of neurons in adulthood could represent a potential protective mechanism. There is also widespread evidence that NF-κB is activated in cerebral ischemia, mainly in neurons, but also in astrocytes, microglia and endothelial cells. Recently, Ridder and Schwaninger [116] comprehensively reviewed the mechanisms of NF-κB activation in cerebral ischemia and discussed the complex signaling network involved in the pathophysiological events taking place in the ischemic brain. In cerebral ischemia it appears that NF-κB activation, with specific involvement of p50-p65 dimers, mainly contributes to neuronal cell death, at least if the insult is severe enough to lead to irreversible brain damage [116].

Overall, both adult neurogenesis and NF-κB signaling appear to be disregulated in a broad range of neurological diseases and disorders. The contribution and significance of this modulation to the etiology and pathogenesis of neurological diseases remain mostly unknown. A link, if any, between the two phenomena also needs further investigation. Certainly the occurrence of adult neurogenesis and its disregulation in many neurodegenerative conditions is intriguing. These findings potentially indicate the endogenous capacity for repair/protective mechanisms in the mammalian brain, and, although this spontaneous regeneration is insufficient to induce neurological improvement, therapeutic implications could be envisioned [117-120].

ADULT NEUROGENESIS AND NF-κB

In 2005 our group demonstrated for the first time the expression of the NF-κB family members in zones of active neurogenesis in the postnatal and adult mouse brain [121]. Since then, Young and collaborators [122] proposed p65-p50 dimers as being required for the normal growth and expansion of neurosphere cultures from mouse embryos. In line with this observation, downregulation of NF-κB activity by overexpression of mutated IκBα repressed proliferation of immortalized neural progenitor cells [123].

Several signals were identified as being able to affect NSC *via* NF-κB activation. The cytokine Tumour Necrosis Factor α (TNF-α) was identified as an *in vitro* inducer of adult rat neural stem cell proliferation *via* NF-κB, as confirmed by pharmacological blockade with a transdominant negative superrepressor IκBα-dn and by IKKβ knock-down through small inhibitory RNA (siRNA) treatment [124]. Erythropoietin (EPO) is a well characterized differentiation factor for late determined and differentiated progenitor cells of hematopoiesis [125]. A study of Shingo *et al.* [126], demonstrated enhanced proliferation of neural progenitors *in vitro* and *in vivo* after EPO treatment. Interestingly, the authors provided evidence that EPO might be a homeostatic autocrine/paracrine signaling molecule that would direct multipotent NSC to become neural progenitors by activating nuclear translocation of NF-κB. More recently, Wada and colleagues [127] demonstrated that Vascular Endothelial Growth Factor (VEGF) *via* its flk1 receptor directly promotes adult NSC survival and that these effects are mediated by NF-κB. In the nervous system glutamate is a well-described activator of NF-κB [15] and Brazel *et al.* [128] showed that glutamate enhances survival and triggers proliferation of SVZ-derived NSC; however glutamate is a pleiotropic mediator controlling many NF-κB independent pathways as well.

In 2007 the group of Michal Schwartz demonstrated that Toll-like receptors (TLR) are expressed by adult neural stem/progenitor cells where they play distinct and opposite functions in NSC proliferation and differentiation both *in vitro* and *in vivo*. Indeed TLR2 deficiency in mice impaired hippocampal neurogenesis whereas the absence of TLR4 resulted in enhanced proliferation and neuronal differentiation. The activation of TLRs on NSCs was mediated *via* MyD88 and Protein Kinase C (PKC) α/β-dependent activation of the NF-κB signaling pathway [129].

More recently, Koo and Duman [130] have demonstrated an essential role of the proinflammatory cytokine Interleukin-1 beta (IL-1β) in the antineurogenic effects of chronic stress. By *in vivo* and *in vitro* studies these authors provided evidence that hippocampal progenitor cells do express IL-1β receptor and that its activation decreases cell proliferation *via* the NF-κB signaling pathway. Indeed, inhibitors of NF-κB/IKK signaling (such as JSH-23 and SC-514, respectively) significantly blocked the antineurogenic effects of IL-1β in adult hippocampal progenitors [130].

NF-κB proteins may be involved not only in neural stem/progenitor cell proliferation and differentiation but also migration. For example, the chemokine monocyte chemoattractant protein-1 (MCP-1) is an NF-κB target gene and stimulates migration of adult rat NSC through the chemokine (C-C motif) receptor 2 (CCR2) [131]. Interestingly, upregulation of MCP-1 in ischemia correlates with the activation of NF-κB [132].

We recently decided to investigate adult hippocampal neurogenesis in p50$^{-/-}$ mice [61]. We could demonstrate through 5-bromo-2-deoxyuridine (BrdU) labelling and Ki67 antigen expression that the rate of proliferation within the DG was not significantly different between p50$^{-/-}$ and wt littermates. Conversely, the late survival (21 days) of newborn BrdU$^+$ cells was drastically reduced in mutant mice. Analysis of the location of BrdU$^+$ cells in the different subregions of the dentate gyrus showed that there was a statistically significant reduction in the number of BrdU$^+$ cells in the granular cell layer and in the hilus but not in the subgranular zone, possibly suggesting a more pronounced effect of p50 absence in regions where newly generated cells migrate and differentiate rather than in the SGZ, where undifferentiated progenitors are located. When lineage fate of newly generated cells was evaluated, the same fraction of BrdU$^+$ cells coexpressed the neuronal marker NeuN in both genotypes, suggesting that overall cell fate is not altered in the mutant mice, although, in absolute numbers, they harbored significantly fewer new neurons than wt mice. When we phenotipically characterized the cell populations affected by the absence of NF-κB p50, we observed no difference in the number of doublecortin (DCX)-positive neuroblasts but a marked reduction of calretinin (CR)-positive postmitotic neurons in the DG of mutant mice. Altogether, these studies suggested highly selective defects in adult neurogenesis in the absence of p50 protein, possibly occurring at the transition between the maturation stages of newly born neuroblasts/neurons marked by the expression of DCX and CR. We provided evidence that the marked reduction in late survival of newborn BrdU$^+$ cells in p50$^{-/-}$ mice could not be attributed to increased apoptosis, as measured by Terminal deoxynucleotidyl transferase dUTP nick end labeling (TUNEL) staining, although we cannot exclude that apoptosis could have occurred at an earlier time point. The homeostatic mechanisms that regulate neuron addition to neuronal networks in adulthood are largely unknown. It is not clear whether in the hippocampus newly generated neurons may be preferentially substituted by more recently generated neurons or if addition of new neurons is in part balanced by loss of old neurons [52, 133]. Given the complex role of NF-κB family members in regulating cell death and/or survival [36, 134, 27], one possibility, to be experimentally addressed, is that in p50$^{-/-}$ mice the minor insertion of new neurons into preexisting circuits may be counterbalanced by a reduced elimination rate of older mature neurons. To test the physiological consequences of the observed alterations in hippocampal neurogenesis, wt and p50$^{-/-}$ mice were evaluated behaviorally. When tested in the Morris water maze, wt and p50$^{-/-}$ mice performed equally in the acquisition test. Moreover in the probe test, performed 24 h after the last training trial, all mice showed a target quadrant preference and the increased time spent in the target quadrant in search of the missing platform suggested normal retrieval of hippocampal-dependent spatial memory in both p50$^{-/-}$ and wt mice. Mice were then subjected to the place recognition test that evaluates hippocampal-dependent short-term spatial memory by testing their ability to discriminate a familiar versus a novel environment. Compared with wt mice, p50$^{-/-}$ mice showed a selective impairment in short-term spatial memory. In summary, p50$^{-/-}$ mice not only exhibited specific deficits in net adult hippocampal neurogenesis but also a specific impairment in a hippocampal-dependent task of short-term spatial memory. These data do not imply that a cause-effect relationship between neurogenesis defects and selective deficits in short-term spatial memory exists in p50$^{-/-}$ mice, but certainly the correlation between these phenomena would deserve further investigation. Interestingly, we have recently evaluated if any alteration was present, in absence of p50, in another adult neurogenic region, the SVZ. No defect could be observed in the SVZ and in the OB of p50$^{-/-}$ compared to wt mice, revealing a hippocampus-specific influence of this NF-κB family member (M. Grilli *et al.*, *unpublished data*).

The role of NF-κB in regulating expression of hippocampal genes involved in selection of newly generated neurons or in synaptic plasticity should be further pursued. Young granule neurons differ substantially from neighboring mature granule cells because they exhibit special properties in synaptic plasticity, such as a lower threshold for the induction of long-term potentiation [135-136]. This enhanced synaptic plasticity appears to be present in cells at the late DCX phase, which means at the transition between the maturation stages that are altered in the p50$^{-/-}$ mice. Incidentally, such a property seems to be very important for some forms of hippocampal-dependent learning and memory [137]. As previously mentioned, mice deficient for other NF-κB subunits have been shown to exhibit deficits in specific hippocampal dependent cognitive tasks [27, 33, 35], although no correlation has ever been attempted with adult neurogenesis. It would certainly be of great interest to investigate whether the deficits in cognitive performance of other mouse lines genetically impaired in the NF-κB signaling would present abnormalities in hippocampal neurogenesis. Since the *bonafide* neural stem cell expresses GFAP, cognitive abnormalities in female GFAP-IκBα-dn mice should also be correlated with any sex-specific disruption of hippocampal neurogenesis [37].

POTENTIAL MOLECULAR PARTICIPANTS LINKING NF-κB SIGNALING, NEUROGENESIS AND COGNITION

At present, the molecular targets of the NF-κB cascade in memory formation and neurogenesis, are largely unknown since only a few studies have attempted to study NF-κB target genes in the context of memory formation.

By microarray analysis, Kaltschmidt and colleagues [35] demonstrated a dramatically reduced expression of the α catalytic subunit of protein kinase A (PKA) in tTA/IκBα-dn double transgenic mice compared to doxycycline-treated animals or to single IκBα-dn mice. The same group could demonstrate that the PKA-dependent pathway of CREB phosphorylation is repressed in NF-κB forebrain deficient mice. Since PKA and CREB have been proposed as critical regulators of long-term memory and LTP [138-140] these results suggest a novel transcriptional signaling cascade in neurons, in which NF-κB regulates the PKA/CREB pathway to function in learning and memory. Notably, PKA also regulates the phosphorylation and activity of NF-κB itself.

Bracchi-Ricard [37] investigated the hippocampal and cortical gene expression profile in GFAP-IκBα-dn female (cognitively impaired) vs male (not impaired) mice. The behavioral deficits in female transgenic mice were correlated with reduced expression of metabotropic glutamate receptor mGluR5 in hippocampus and cortex. These data are of interest based on several literature data. Indeed gene targeting experiments by Jia and colleagues revealed a role for the glutamate receptor mGluR5 in learning and memory [141]. O'Riordan *et al.* [142] provided evidence that the NF-κB family is regulated by Group I (GpI) mGluRs in the hippocampus and that c-Rel is necessary for formation of long-term memory. Additionally, Pizzi *et al.* [89] suggested that regulation of gene expression by GpI-mGluRs in cortical neurons relies on activation of c-Rel. Interestingly, mGluR5 are also expressed in areas of adult neurogenesis [143].

Similar to other neural activity-induced plasticity, adult neurogenesis is modulated by a plethora of external stimuli. For example, synchronized activation of mature dentate neurons by electroconvulsive treatment (ECT) in adult mice causes sustained up-regulation of hippocampal neurogenesis. How transient activation of mature neuronal circuits modulates adult neurogenesis over days and weeks is largely unknown. Gadd45 is a NF-κB-dependent anti-apoptotic gene [144] which was recently identified as a neural activity-induced immediate early gene which regulates adult hippocampal neurogenesis. Mice with a Gadd45b deletion exhibited specific deficits in neural activity-induced proliferation of neural progenitors and dendritic growth of newborn neurons in the adult hippocampus. More specifically, Gadd45b was required for activity-induced DNA demethylation of specific promoters and expression of corresponding genes critical for adult neurogenesis, including brain-derived neurotrophic factor (BDNF) and FGF genes [145]. Overall these findings suggest a model in which Gadd45b links neuronal circuit activity to epigenetic DNA modification and expression of secreted factors for extrinsic modulation of neurogenesis in the adult brain.

Another potential pathway which may link NF-κB-mediated signaling with neurogenesis and possibly cognitive impairment is the one involved in the APP metabolism. The proteolytic processing of APP has long been studied because of its association with the pathology of AD. The ectodomain of APP is indeed shed by alpha- or beta-secretase cleavage. The remaining membrane bound stub can then undergo regulated intramembrane proteolysis by gamma-secretase, releasing Aβ which accumulates in the brain of AD patients. In addition, alpha- or beta-secretase cleavage releases the APP intracellular domain (AICD). The physiological functions of this proteolytic processing are not well understood but recently Ma *et al.* [146] suggested that TAG1 is a functional ligand for APP and that interaction between TAG1 and APP triggers gamma-secretase-dependent release of AICD which in turn interacts with Fe65. The AICD/Fe65 complex has been shown to display transcriptional activity [147-148]. Interestingly,

TAG1, APP and Fe65, colocalized in the adult neural stem cell niche of the subventricular zone and in fetal neural progenitor cells *in vitro*. Experiments in TAG1, APP and Fe65 null mice as well as TAG1 and APP double-null mice demonstrate that TAG1-APP signaling induces a gamma-secretase- and Fe65-dependent suppression of neurogenesis [146]. Interestingly, this intracellular signal transduction pathway is influenced by the NF-κB signaling pathway. Activation of the NF-κB pathway, for example by overexpression of NIK, downregulates the transcriptional activity of the AICD-Fe65 complex [149]. Whether this pathway is also operative in the hippocampus and may contribute to disregulated neurogenesis in AD or other neurodegenerative disorders needs to be further investigated. Another interesting piece of data comes from Baek *et al.* [150], who have demonstrated that exchange of N-CoR Corepressor and Tip60 Coactivator Complexes Links Gene Expression by NF-κB and APP. They showed that IL-1β caused nuclear export of a specific N-CoR corepressor complex, resulting in derepression of a specific subset of NF-kappaB-regulated genes, among which tetraspanin KAI1, a protein that regulates membrane receptor function. Moreover, nuclear export of the N-CoR/TAB2/HDAC3 complex by IL-1β was temporally linked to selective recruitment of a Tip60 coactivator complex. Interestingly, KAI1 was directly activated by the AICD, Fe65, and Tip60 ternary complex, identifying a specific *in vivo* gene target of an APP-dependent transcription complex in the brain. In cell lines, the continued presence of p50, but not p65, on the *KAI1* promoter by chromatin immunoprecipitation (ChIP) assay was demonstrated indicating that this promoter might recruit p50 homodimers. Bcl$_3$ was also detected on the *KAI1* promoter, further supporting the idea that this promoter recruits a p50 homodimer that binds both Bcl$_3$ and a specific N-CoR corepressor complex when it is transcriptionally inactive. Again, whether these nuclear corepressor/coactivator exchange events may also take place in NSC may be of interest. Additionally these studies may possibly help identifying subsets of NF-κB regulated genes involved in modulation of neurogenesis.

Also the soluble derivative of APP metabolism (sAPP) may be linked to neurogenesis. Caillé *et al.* [151] showed that the SVZ is a major sAPP binding site and that binding occurs on progenitor cells expressing the EGF receptor. They also demonstrated a new function for sAPP as a regulator of SVZ progenitor proliferation in the adult central nervous system. Indeed *in vivo*, sAPP infusions increased the number of Epidermal Growth Factor (EGF)-responsive progenitors through their increased proliferation. Conversely, blocking sAPP secretion or downregulating APP synthesis decreased proliferation of EGF-responsive cells, which in turn led to a reduction of the pool of progenitors. Interestingly, APP gene expression is also under NF-κB regulation, and in particular of p50 homodimers [152].

CONCLUSIONS

A vast array of information suggests that NF-κB signaling is involved in the pathogenesis of neurological disorders such as AD, PD, HD, stroke, brain trauma, either promoting or mitigating disease [29, 36, 116, 134, 153]. Molecular pathways upstream and downstream of NF-κB in the CNS have been under investigation due to the potential to provide novel targets for therapeutic intervention. Several neurodegenerative disorders characterized by disregulated NF-κB transcription have also been associated with alterations of neurogenesis [117-120], but no attempt has been made to correlate these two phenomena. Whether or not alterations in NF-κB-mediated signaling may contribute to disregulated neurogenesis described in such disorders remains to be unambiguously confirmed. Additionally, alterations in NF-κB signaling have been reported during aging in the hippocampus [154]. It would be of interest to evaluate whether these alterations may also potentially contribute to age-dependent changes in the neurogenic potential of the mammalian brain and/or to age-related cognitive impairment.

Our current understanding of adult neurogenesis remains limited. Despite this, pharmacological modulation of adult neurogenesis holds the potential for being a strategy to be explored for treating neurological and psychiatric diseases that have been resistant to conventional medications or with no current treatment. There is the possibility that neurogenesis may indeed contribute to the regeneration of neuronal circuitry under pathological conditions, or that it may be associated with their symptoms and/or pathophysiology. For the development of successful neuroregenerative therapies, it is therefore crucial to comprehensively understand the mechanisms that regulate neurogenesis in both physiological and pathological conditions.

The novel finding of a functional involvement of NF-κB in adult neurogenesis adds further complexity to the potential contribution of this family of transcription factors to neurological disorders. For example, in disorders such as stroke NF-κB signaling is generally considered detrimental [116]. On the other hand enhanced neurogenesis has been documented in human and experimental stroke [111] and suggested to represent an endogenous attempt to regenerate cells lost as a consequence of the insult [155]. Future research aimed at developing novel NF-κB-based preventative and therapeutic approaches in neurological disorders should take into consideration the possibility that NF-κB may also be involved in regulating a potentially protective response such as neurogenesis. This would imply the need to search for therapeutic agents that may inhibit NF-κB-mediated detrimental and maladaptive effects without compromising the potentially adaptive ones.

ACKNOWLEDGEMENTS

The authors would like to apologize with all researchers whose work could not be quoted, due to space limitation. M.G. was supported by grants from Fondazione Cariplo, Fondazione delle Comunità del Novarese and by the Italian Ministero dell'Istruzione, Università e Ricerca (MIUR), under the Progetti di Interesse Nazionale (PRIN) framework. The authors declare the absence of any potential conflict of interest regarding this chapter.

ABBREVIATIONS

ACh	=	Acetylcholine
AD	=	Alzheimer's disease
Aβ	=	Amyloid beta
APP	=	Amyloid Precursor Protein
AICD	=	APP intracellular domain
BAFF	=	TNF Family Member B Cell-Activating Factor
BDNF	=	Brain-derived neurotrophic factor
BrdU	=	5-bromo-2-deoxyuridine
CamKII	=	Ca^{2+}/calmodulin-dependent protein kinase II
CR	=	Calretinin
CCR2	=	Chemokine (C-C motif) receptor
ChEI	=	Cholinesterase inhibitors
ChIP	=	Chromatin immunoprecipitation
DG	=	Dentate gyrus
dn	=	Dominant negative
DCX	=	Doublecortin

ECT	=	Electroconvulsive treatment
EGF	=	Epidermal growth factor
EE	=	Environmental enrichment
EPO	=	Erythropoietin
FGF	=	Fibroblast growth factor
Gadd45	=	Growth Arrest and DNA Damage-inducible gene
GCL	=	Granular cell layer
GFAP	=	Glial fibrillary acidic protein
GpI mGluRs	=	Group I metabotropic glutamate receptors
Htt	=	Hungtingtin
HD	=	Hungtington's disease
IKK	=	IκB Kinase
IκB	=	Inhibitor of NF-κB
IL-1β	=	Interleukin-1 β
IL-6	=	Interleukin-6
IL-8	=	Interleukin-8
LPS	=	Lipopolysaccaride
LTP	=	Long term potentiation
LTD	=	Long term depression
M-CSF	=	Macrophage-colony stimulating factor
MnSOD	=	Manganese superoxide dismutase
mGluR5	=	Metabotropic glutamate receptor 5
MPTP	=	1-methyl-4-phenyl-1,2,3,6-tetrahydropteridine
MCAO	=	Middle cerebral artery occlusion
MCP-1	=	Monocyte chemoattractant protein-1
MyD88	=	Myeloid differentiation primary response gene (88)
NeuN	=	Neuronal nuclei antigen

NIK	=	NF-κB inducing kinase
NPC	=	Neural progenitor cells
NSC	=	Neural stem cells
NOS-II	=	Nitric oxide synthase 2
6-OHDA	=	6-hydroxydopamine
OB	=	Olfactory bulb
PD	=	Parkinson's disease
3PN	=	3-nitropropionic acid
PSD95	=	Postsynaptic density protein 95
PS1	=	Presenilin 1
PKA	=	Protein kinase A
PKCα/β	=	Protein kinase C α/β
RHD	=	Rel homology domain
RAGE	=	Receptor for advanced glycation end-products
siRNA	=	Small inhibitory RNA
sAPP	=	Soluble amyloid precursor protein
sAPPα	=	Soluble amyloid precursor protein α
SGZ	=	SubGranular Zone
SVZ	=	SubVentricular Zone
TUNEL	=	Terminal deoxynucleotidyl transferase dUTP nick end labeling
tTA	=	Tetracycline transactivator
TLR	=	Toll-like receptor
TAD	=	Transactivator domains
TNF	=	Tumor necrosis factor
TNFR	=	Tumor necrosis factor receptor
VEGF	=	Vascular endothelial growth factor
wt	=	wild-type

REFERENCES

[1] Sen R, Baltimore D. Inducibility of kappa immunoglobulin enhancer-binding protein NF-kappa B by a post translational mechanism. Cell 1986; 47: 921-8.

[2] Krakauer T. Nuclear factor-kappaB: fine-tuning a central integrator of diverse biologic stimuli. Int Rev Immunol 2008; 27: 286-92.

[3] Gerondakis S, Grumont R, Gugasyna R, *et al.* Unravelling the complexities of the NF-kappaB signaling pathway using mouse knockout and transgenic models. Oncogene 2006; 25: 6781-99.

[4] Grilli M, Chiu JJ-S, Lenardo MJ. NF-κB and Rel: participants in a multiform transcriptional regulatory system. Int Rev Cytol 1993; 143: 01-62.

[5] Pizzi M, Spano PF. Distinct roles of diverse nuclear factor-kappaB complexes in neuropathological mechanisms. Eur J Pharmacol 2006; 545: 22-8.

[6] Leung TH, Hoffmann A, Baltimore D. One nucleotide in a kappaB site can determine cofactor specificity for NF-kappaB dimmers. Cell 2004; 118: 453-64.

[7] Bates PW, Miyamoto S. Expanded nuclear roles for IκBs. Sci STKE 2004; 254: 01-2.

[8] Van Loo G, De Lorenzi R, Schmidt H, *et al.* Inhibition of transcription factor NF-kappaB in the central nervous system ameliorates autoimmune encephalomyelitis in mice. Nat Immunol 2006; 7: 954-61.

[9] Huang TT, Wuerzberger-Davis SM, Wu ZH, Miyamoto S. Sequential modification of NEMO/IKKgamma by SUMO-1 and ubiquitin mediates NF-kappaB activation by genotoxic stress. Cell 2003; 115: 565-76.

[10] Saccani S, Marazzi I, Beg AA, Natoli G. Degradation of promoter-bound p65/RelA is essential for the prompt termination of the nuclear factor kappa B response. J Exp Med 2004; 200: 107-13.

[11] Chen LF, Williams SA, Mu Y, *et al.* NF-kappaB RelA phosphorylation regulates RelA acetylation. Mol Cell Biol 2005; 25: 7966-75.

[12] Franzoso G, Bours V, Park S, Tomita-Yamaguchi M, Kelly K, Siebenlist U. The candidate oncoprotein Bcl₃ is an antagonist of p50/NF-kappaB-mediated inhibition. Nature 1992; 359: 339-42.

[13] Bours V, Franzoso G, Azarenko V, *et al.* The oncoprotein BCL-3 directly transactivates through kappa B motifs via association with DNA-binding p50 homodimers. Cell 1993; 72: 729-39.

[14] Kaltschmidt C, Kaltschmidt B, Neumann H, Wekerle H, Baeuerle PA. Constitutive NF-kappaB activity in neurons. Mol Cell Biol 1994; 14: 3981-92.

[15] Guerrini L, Blasi F, Denis-Donini S. Synaptic activation of NF-kappaB by glutamate in cerebellar granule neurons *in vitro*. Proc Natl Acad Sci USA 1995; 92: 9077-81.

[16] Meberg PJ, Kinney WR, Valcourt EG, Routtenberg A. Gene expression of the transcription factor NF-kappaB in hippocampus: regulation by synaptic activity. Brain Res Mol Brain Res 1996; 38: 179-90.

[17] Wellmann H, Kaltschmidt B, Kaltschmidt C. Retrograde transport of transcription factor NF-kappaB in living neurons. J Biol Chem 2001; 276: 11821-9.

[18] Meffert MK, Chang JM, Wiltgen BJ, Fanselow MS, Baltimore D. NF-kappaB functions in synaptic signaling and behavior. Nat Neurosci 2003; 6: 1072-8.

[19] Schölzke MN, Potrovita I, Subramaniam S, Prinz S, Schwaninger M. Glutamate activates NF-kappaB through calpain in neurons. Eur J Neurosci 2003; 8: 3305-10.

[20] Suzuki T, Mitake S, Okumura-Noji K, Yang JP, Fujii T, Okamoto T. Presence of NF-kappaB-like and IkappaB-like immunoreactivities in postsynaptic densities. Neuroreport 1997; 8: 2931-5.

[21] Albensi BC, Mattson MP. Evidence for the involvement of TNF and NF-kappaB in hippocampal synaptic plasticity. Synapse 2000; 35: 151-9.

[22] Mattson MP, Culmsee C, Yu Z, Camandola S. Roles of nuclear factor kappa B in neuronal survival and plasticity. J Neurochem 2000; 74: 443-56.

[23] Burr PB, Morris BJ. Involvement of NMDA receptors and a p21Ras-like guanosine triphosphatase in the constitutive activation of nuclear factor-kappaB in cortical neurons. Exp Brain Res 2002; 147: 273-9.

[24] Lilienbaum A, Israël A. From calcium to NF-kappaB signaling pathways in neurons. Mol Cell Biol 2003; 23: 2680-98.

[25] Freudenthal R, Romano A, Routtenberg A. Transcription factor NF-kappaB activation after *in vivo* perforant path LTP in mouse hippocampus. Hippocampus 2004; 14: 677-83.

[26] Yeh SH, Lin CH, Lee CF, Gean PW. A requirement of nuclear factor-kappaB activation in fear-potentiated startle. J Biol Chem 2002; 277: 46720-9.

[27] Meffert MK, Baltimore D. Physiological functions for brain NF-κB. TINS 2004; 28: 37-43.

[28] Mattson MP. NF-kappaB in the survival and plasticity of neurons. Neurochem Res 2005; 30: 883-93.

[29] Memet S. NF-κB functions in the nervous system: from development to disease. Biochem Pharmacol 2006; 72: 1180-95.

[30] Romano A, Freudenthal R, Merlo E, Routtenberg A. Evolutionary conserved role of the NF-κB transcription factor in neural plasticity and memory. Eur J Neurosci 2006; 24: 1507-16.

[31] Beg AA, Sha WC, Bronson RT, Ghosh S, Baltimore D. Embryonic lethality and liver degeneration in mice lacking the RelA component of NF-kappaB. Nature 1995; 376: 167-70.

[32] Alcamo E, Mizgerd JP, Horwitz BH, *et al.* Targeted mutation of TNF receptor I rescues the RelA-deficient mouse and reveals a critical role for NF-kappaB in leukocyte recruitment. J Immunol 2001; 167: 1592-00.

[33] Levenson JM, Choi S, Lee SY, *et al.* A bioinformatics analysis of memory consolidation reveals involvement of the transcription factor c-rel. J Neurosci 2004; 24: 3933-43.

[34] Ahn HJ, Hernandez CM, Levenson JM, Lubinm FD, Liou HC, Sweatt JD. c-Rel, an NF-kappaB family transcription factor, is required for hippocampal long-term synaptic plasticity and memory formation. Learn Mem 2008; 15: 539-49.

[35] Kaltschmidt B, Ndiaye D, Korte M, *et al.* NF-kappaB regulates spatial memory formation and synaptic plasticity through protein kinase A/CREB signaling. Mol Cell Biol 2006; 26: 2936-46.

[36] Grilli M, Memo M. Nuclear factor-kappaB/Rel proteins: a point of convergence of signalling pathways relevant in neuronal function and dysfunction. Biochem Pharmacol 1999; 57: 01-7.

[37] Bracchi-Ricard V, Brambilla R, Levenson J, *et al.* Astroglial nuclear factor-kappaB regulates learning and memory and synaptic plasticity in female mice. J Neurochem 2008; 104: 611-23.

[38] Sha WC, Liou HC, Tuomanen EI, Baltimore D. Targeted disruption of the p50 subunit of NF-kappa B leads to multifocal defects in immune responses. Cell 1995; 80: 321-30.

[39]   Kassed CA, Willing AE, Garbuzova-Davis S, Sanberg PR, Pennypacker KR. Lack of NF-kappaB p50 exacerbates degeneration of hippocampal neurons after chemical exposure and impairs learning. Exp Neurol 2002; 176: 277-88.

[40]   Kassed CA, Herkenham M. NF-kappaB p50-deficient mice show reduced anxiety-like behaviors in tests of exploratory drive and anxiety. Behav Brain Res 2004; 154: 577-84.

[41]   Ashburner BP, Westerheide SD, Baldwin AS Jr. The p65 (RelA) subunit of NF-kappaB interacts with the histone deacetylase (HDAC) corepressors HDAC1 and HDAC2 to negatively regulate gene expression. Mol Cell Biol 2001; 21: 7065-77.

[42]   Viatour P, Legrand-Poels S, van Lint C, *et al.* Cytoplasmic IkappaBalpha increases NF-kappaB-independent transcription through binding to histone deacetylase (HDAC) 1 and HDAC3. J Biol Chem 2003; 278: 46541-8.

[43]   Anest V, Cogswell PC, Baldwin AS Jr. IkappaB kinase alpha and p65/RelA contribute to optimal epidermal growth factor-induced c-fos gene expression independent of IkappaBalpha degradation. J Biol Chem 2004; 279: 31183-9.

[44]   Colvis CM, Pollock JD, Goodman RH, *et al.* Epigenetic mechanisms and gene networks in the nervous system. J Neurosci 2005; 25: 10379-89.

[45]   Levenson JM, Sweatt JD. Epigenetic mechanisms in memory formation. Nat Rev Neurosci 2005; 6: 108-18.

[46]   Lubin FD, Sweatt JD. The IkappaB kinase regulates chromatin structure during reconsolidation of conditioned fear memories. Neuron 2007; 55: 942-57.

[47]   Seri B, Garcia-Verdugo JM, McEwen BS, Alvarez-Buylla A. Astrocytes give rise to new neurons in the adult mammalian hippocampus. J Neurosci 2001; 21: 7153-60.

[48]   Kempermann G, Jessberger S, Steiner B, Kronenberg G. Milestones of neuronal development in the adult hippocampus. Trends Neurosci 2004; 27: 447-52.

[49]   Seri B, Garcia-Verdugo JM, Collado-Morente L, McEwen BS, Alvarez-Buylla A. Cell types, lineage, and architecture of the germinal zone in the adult dentate gyrus. J Comp Neurol 2004; 478: 359-78.

[50]   Ming GL, Song H. Adult neurogenesis in the mammalian central nervous system. Annu Rev Neurosci 2005; 28: 223-50.

[51]   Biebl M, Cooper CM, Winkler J, Kuhn HG. Analysis of neurogenesis and programmed cell death reveals a self renewing capacity in the adult rat brain. Neurosci Lett 2000; 291: 17-20.

[52]   Lehmann K, Butz M, Teuchert-Noodt G. Offer and demand: proliferation and survival of neurons in the dentate gyrus. Eur J Neurosci 2005; 21: 3205-16.

[53]   Gould E, Tanapat P, Hastings NB, Shors TJ. Neurogenesis in adulthood: a possible role in learning. Trends Cogn Sci 1999; 3: 186-92.

[54]   van Praag H, Christie BR, Sejnowski TJ, Gage FH. Running enhances neurogenesis, learning, and long-term potentiation in mice. Proc Natl Acad Sci USA 1999; 96: 13427-31.

[55]   Shors TJ, Miesagaes G, Beylin A, Zhao M, Rydel T, Gould E. Neurogenesis in the adult is involved in the formation of trace memories. Nature 2001; 410: 372-76.

[56]   Kempermann G, Gage FH. Genetic determinants of adult hippocampal neurogenesis correlate with acquisition, but not probe trial performance, in the water maze task. Eur J Neurosci 2002; 16: 129-36.

[57]   Aimone JB, Wiles J, Gage FH. Potential role for adult neurogenesis in the encoding of time in new memories. Nat Neurosci 2006; 9: 723-27.

[58]   Leuner B, Gould E, Shors TJ. Is there a link between adult neurogenesis and learning? Hippocampus 2006; 16: 216-24.

[59]   Kee N, Teixairan CM, Wang AH, Frankland PW. Preferential incorporation of adult-generated granule cells into spatial memory networks in the dentate gyrus. Nat Neurosci 2007; 10: 355-62.

[60]   Saxe MD, Malleret G, Vronskaya S, *et al.* Paradoxical influence of hippocampal neurogenesis on working memory. Proc Natl Acad Sci USA 2007; 10: 4642-6.

[61]   Denis-Donini S, Dellarole A, Crociara P, *et al.* Impaired adult neurogenesis associated with short-term memory defects in NF-kappaB p50-deficient mice. J Neurosci 2008; 28: 3911-9.

[62]   Tashiro A, Makino H, Gage FH. Experience-specific functional modification of the dentate gyrus through adult neurogenesis: a critical period during an immature stage. J Neurosci 2007; 27: 3252-9.

[63]   Dupret D, Fabre A, Dobrossy MD, *et al.* Spatial learning depends on both the addition and removal of new hippocampal neurons. PLOS Biol 2007; 5: 1683-94.

[64]   Snyder JS, Radik R, Wojtowicz JM, Cameron HA. Anatomical gradients of adult neurogenesis and activity: young neurons in the ventral dentate gyrus are activated by water maze training. Hippocampus 2009; 19: 360-70.

[65]   Brown J, Cooper-Kuhn CM, Kempermann G, *et al.* Enriched environment and physical activity stimulate hippocampal but not olfactory bulb neurogenesis. Eur J Neurosci 2003; 17: 2042-6.

[66]   Bruel-Jungerman E, Laroche S, Rampon C. New neurons in the dentate gyrus are involved in the expression of enhanced long-term memory following environmental enrichment. Eur J Neurosci 2005; 21: 513-21.

[67]   Ueda S, Sakakibara S, Yoshimoto K. Effect of long-lasting serotonin depletion on environmental enrichment-induced neurogenesis in adult rat hippocampus and spatial learning. Neuroscience 2005; 135: 395-02.

[68]   Leal-Galicia P, Castañeda-Bueno M, Quiroz-Baez R, Arias C. Long-term exposure to environmental enrichment since youth prevents recognition memory decline and increases synaptic plasticity markers in aging. Neurobiol Learn Mem 2008; 90: 511-8.

[69]   Gould E, McEwen BS, Tanapat P, Galea LA, Fuchs E. Neurogenesis in the dentate gyrus of the adult tree shrew is regulated by psychosocial stress and NMDA receptor activation. J Neurosci 1997; 17: 2492-8.

[70]   Lemaire V, Koehl M, Le Moal M, Abrous DN. Prenatal stress produces learning deficits associated with an inhibition of neurogenesis in the hippocampus. Proc Natl Acad Sci USA 2000; 97: 11032-7.

[71]   Coe CL, Kramer M, Czéh B, *et al.* Prenatal stress diminishes neurogenesis in the dentate gyrus of juvenile rhesus monkeys. Biol Psychiatry 2003; 54: 1025-34.

[72]   Simon M, Czéh B, Fuchs E. Age-dependent susceptibility of adult hippocampal cell proliferation to chronic psychosocial stress. Brain Res 2005; 1049: 244-8.

[73]   Thomas RM, Hotsenpiller G, Peterson DA. Acute psychosocial stress reduces cell survival in adult hippocampal neurogenesis without altering proliferation. J Neurosci 2007; 27: 2734-43.

[74]   Klempin F, Kempermann G. Adult hippocampal neurogenesis and aging. Eur Arch Psychiatry Clin Neurosci 2007; 257: 271-80.

[75]   Jin K, Peel AL, Mao XO, *et al.* Increased hippocampal neurogenesis in Alzheimer's disease. Proc Natl Acad Sci USA 2004; 6: 343-7.

[76]   Boekhoorn K, Joels M, Lucassen PJ. Increased proliferation reflects glial and vascular-associated changes, but not neurogenesis in the presenile Alzheimer hippocampus. Neurobiol Dis 2006; 24: 01-14.

[77]   Haughey NJ, Liu D, Nath A, Borchard AC, Mattson MP. Disruption of neurogenesis in the subventricular zone of adult mice, and in human cortical neuronal precursor cells in culture, by amyloid beta-peptide: implications for the pathogenesis of Alzheimer's disease. Neuromolecular Med 2002; 1: 125-35.

[78]   Haughey NJ, Nath A, Chan SL, Borchard AC, Rao MS, Mattson MP. Disruption of neurogenesis by amyloid beta-peptide, and perturbed neural progenitor cell homeostasis, in models of Alzheimer's disease. J Neurochem 2002; 83: 1509-24.

[79]   Kaneko N, Okano H, Sawamoto K. Role of the cholinergic system in regulating survival of newborn neurons in the adult mouse dentate gyrus and olfactory bulb. Genes Cells 2006; 1: 1145-59.

[80]   Jankowsky JL, Melnikova T, Fadale DJ, et al. Environmental enrichment mitigates cognitive deficits in a mouse model of Alzheimer's disease. J Neurosci 2005; 25: 5217-24.

[81]   Kaltschmidt B, Uherek M, Volk B, Baeuerle PA, Kaltschmidt C. Transcription factor NF-kappaB is activated in primary neurons by amyloid beta peptides and in neurons surrounding early plaques from patients with Alzheimer disease. Proc Natl Acad Sci USA 1997; 94: 2642-7.

[82]   Terai K, Matsuo A, McGeer PL. Enhancement of immunoreactivity for NF-kappa B in the hippocampal formation and cerebral cortex of Alzheimer's disease. Brain Res 1996; 735: 159-68.

[83]   Boissière F, Hunot S, Faucheux B, et al. Nuclear translocation of NF-kappaB in cholinergic neurons of patients with Alzheimer's disease. Neuroreport 1997; 8: 2849-52.

[84]   Kaltschmidt B, Uherek M, Wellmann H, Volk B, Kaltschmidt C. Inhibition of NF-kappaB potentiates amyloid beta-mediated neuronal apoptosis. Proc Natl Acad Sci USA 1999; 96: 9409-14.

[85]   Behl C, Davis JB, Lesley R, Schubert D. Hydrogen peroxide mediates amyloid beta protein toxicity. Cell 1994; 77: 817-27.

[86]   Du Yan S, Zhu H, Fu J, et al. Amyloid-beta peptide-receptor for advanced glycation endproduct interaction elicits neuronal expression of macrophage-colony stimulating factor: a proinflammatory pathway in Alzheimer disease. Proc Natl Acad Sci USA 1997; 94: 5296-01.

[87]   Valerio A, Boroni F, Benarese M, et al. NF-kappaB pathway: a target for preventing beta-amyloid (Abeta)-induced neuronal damage and Abeta42 production. Eur J Neurosci 2006; 23: 1711-20.

[88]   Barger SW, Horster D, Furukawa K, Goodman Y, Krieglstein J, Mattson MP. Tumor necrosis factors alpha and beta protect neurons against amyloid beta-peptide toxicity: evidence for involvement of a kB-binding factor and attenuation of peroxide and Ca2+ accumulation. Proc Natl Acad Sci USA 1995; 92: 9328-32.

[89]   Pizzi M, Sarnico I, Boroni F, et al. NF-kappaB factor c-Rel mediates neuroprotection elicited by mGlu5 receptor agonists against amyloid beta-peptide toxicity. Cell Death Differ 2005; 12: 761-72.

[90]   Guo Q, Robinson N, Mattson MP. Secreted beta-amyloid precursor protein counteracts the pro-apoptotic action of mutant presenilin-1 by activation of NF-kB and stabilization of calcium homeostasis. J Biol Chem 1998; 273: 12341-5.

[91]   Walker DG, Lue LF, Beach TG. Gene expression profiling of amyloid beta peptide-stimulated human post-mortem brain microglia. Neurobiol Aging 2001; 22: 957-66.

[92]   Chen J, Zhou Y, Mueller-Steiner S, et al. SIRT1 protects against microglia-dependent amyloid-beta toxicity through inhibiting NF-kappaB signaling. J Biol Chem 2005; 280: 40364-74.

[93]   Curtis MA, Penney EB, Pearson AG. Increased cell proliferation and neurogenesis in the adult human Huntington's disease brain. Proc Natl Acad Sci USA 2003; 100: 9023-7.

[94]   Curtis MA, Penney EB, Pearson J, Dragunow M, Connor B, Faull RL. The distribution of progenitor cells in the subependymal layer of the lateral ventricle in the normal and Huntington's disease human brain. Neuroscience 2005; 132: 777-88.

[95]   Batista CM, Kippin TE, Willaime-Morawek S, Shimabukuro MK, Akamatsu W, van der Kooy D. A progressive and cell non-autonomous increase in striatal neural stem cells in the Huntington's disease R6/2 mouse. J Neurosci 2006; 26: 10452-60.

[96]   Jin K, LaFevre-Bernt M, Sun Y, et al. FGF-2 promotes neurogenesis and neuroprotection and prolongs survival in a transgenic mouse model of Huntington's disease. Proc Natl Acad Sci USA 2005; 102: 18189-94.

[97]   Lazic SE, Grote H, Armstrong RJ, et al. Decreased hippocampal cell proliferation in R6/1 Huntington's mice. Neuroreport 2004; 15: 811-3.

[98]   Gil JM, Mohapel P, Araújo IM, Popovic N, Li JY, Brundin P, Petersén A. Reduced hippocampal neurogenesis in R6/2 transgenic Huntington's disease mice. Neurobiol Dis 2005; 20: 744-51.

[99]   van Dellen A, Cordery PM, Spires TL, Blakemore C, Hannan AJ. Wheel running from a juvenile age delays onset of specific motor deficits but does not alter protein aggregate density in a mouse model of Huntington's disease. BMC Neurosci 2008; 01: 34.

[100]   Khoshnan A, Ko J, Watkin EE, Paige LA, Reinhart PH, Patterson PH. Activation of the IkappaB kinase complex and nuclear factor-kappaB contributes to mutant huntingtin neurotoxicity. J Neurosci 2004; 24: 7999-08.

[101]   Qin ZH, Chen RW, Wang Y, Nakai M, Chuang DM, Chase TN. Nuclear factor kappaB nuclear translocation upregulates c-Myc and p53 expression during NMDA receptor-mediated apoptosis in rat striatum. J Neurosci 1999; 19: 4023-33.

[102]   Yu Z, Zhou D, Cheng G, Mattson MP. Neuroprotective role for the p50 subunit of NF-kappaB in an experimental model of Huntington's disease. J Mol Neurosci 2000; 15: 31-44.

[103]   Höglinger GU, Rizk P, Muriel MP, et al. Dopamine depletion impairs precursor cell proliferation in Parkinson disease. Nat Neurosci 2004; 7: 726-35.

[104]   Hunot S, Brugg B, Ricard D, et al. Nuclear translocation of NF-κB is increased in dopaminergic neurons of patients with parkinson disease. Proc Natl Acad Sci USA 1997; 94: 7531-6.

[105]   Blum D, Torch, S, Nissou MF, Verna JM. 6-Hydroxydopamine-induced nuclear factor-kappa B activation in PC12 cells. Biochem Pharmacol 2001; 62: 473-81.

[106]   Panet H, Barzilai A, Daily D, Melamed E, Offen D. Activation of nuclear transcription factor kappa B (NFkappaB) is essential for dopamine-induced apoptosis in PC12 cells. J Neurochem 2001; 77: 391-8.

[107]   Lee HJ, Kim SH, Kim KW, et al. Anti-apoptotic role of NF-kappaB in the auto-oxidized dopamine-induced apoptosis of PC12 cells. J Neurochem 2001; 76: 602-9.

[108]   Levites Y, Youdim MB, Maor G, Mandel S. Attenuation of 6-hydroxydopamine (6-OHDA)-induced nuclear factor-kappaB (NF-kappaB) activation and cell death by tea extracts in neuronal cultures. Biochem Pharmacol 2002; 63: 21-29.

[109]   Park SH, Choi WS, Yoon SY, Ahn YS, Oh YJ. Activation of NF-kappaB is involved in 6-hydroxydopamine-but not MPP+-induced dopaminergic neuronal cell death: its potential role as a survival determinant. Biochem Biophys Res Commun 2004; 322: 727-33.

[110]   Teismann P, Schwaninger M, Weih F, Ferger B. Nuclear factor-kappaB activation is not involved in a MPTP model of Parkinson's disease. Neuroreport 2001; 12: 1049-53.

[111]   Jin K, Wang X, Xie L, *et al.* Evidence for stroke-induced neurogenesis in the human brain. Proc Natl Acad Sci USA 2006; 103: 13198-02.

[112]   Liu J, Solway K, Messing RO, Sharp FR. Increased neurogenesis in the dentate gyrus after transient global ischemia in gerbils. J Neurosci 1998; 18: 7768-78.

[113]   Nakatomi H, Kuriu T, Okabe S, *et al.* Regeneration of hippocampal pyramidal neurons after ischemic brain injury by recruitment of endogenous neural progenitors. Cell 2002; 110: 429-41.

[114]   Arvidsson A, Kokaia Z, Lindvall O. N-methyl-D-aspartate receptor-mediated increase of neurogenesis in adult rat dentate gyrus following stroke. Eur J Neurosci 2001; 14: 10-8.

[115]   Yamashita T, Ninomiya M, Hernández Acosta P, *et al.* Subventricular zone-derived neuroblasts migrate and differentiate into mature neurons in the post-stroke adult striatum. J Neurosci 2006; 26: 6627-36.

[116]   Ridder DA, Schwaninger M. NF-kappaB signaling in cerebral ischemia. Neuroscience 2009; 158: 995-06.

[117]   Steiner B, Wolf S, Kempermann G. Adult neurogenesis and neurodegenerative disease. Regen Med 2006; 1: 15-28.

[118]   Galvan V, Bredesen DE. Neurogenesis in the adult brain: implications for Alzheimer's disease. CNS Neurol Disord Drug Targets 2007; 6: 303-10.

[119]   Kempermann G, Krebs J, Fabel K. The contribution of failing adult hippocampal neurogenesis to psychiatric disorders. Curr Opin Psychiatry 2008; 21: 290-5.

[120]   Abdipranoto A, Wu S, Stayte S, Vissel B. The role of neurogenesis in neurodegenerative diseases and its implications for therapeutic development. CNS Neurol Disord Drug Targets 2008; 7: 187-10.

[121]   Denis-Donini S, Caprini A, Frassoni C, Grilli M. Members of the NF-kappaB family expressed in zone of active neurogenesis in the postnatal and adult mouse brain. Brain Res Dev Brain Res 2005; 154: 81-9.

[122]   Young KM, Bartlett PF, Coulson EJ. Neural progenitor number is regulated by nuclear factor-kappaB p65 and p50 subunit-dependent proliferation rather than cell survival. J Neurosci Res 2006; 83: 39-49.

[123]   Zhu C, Liu Z, Gui L, Yao W, Qian W, Zhang C. Mutated IkappaBalpha represses proliferation of immortalized neural progenitor cells and prevents their apoptosis after oxygen-glucose deprivation. Brain Res 2008; 1244: 24-31.

[124]   Widera D, Mikenberg I, Elvers M, Kaltschmidt C, Kaltschmidt B. Tumor necrosis factor alpha triggers proliferation of adult neural stem cells via IKK/NF-kappaB signaling. BMC Neurosci 2006; 7: 64.

[125]   Geissler D, Konwalinka G, Peschel C, Braunsteiner H. The role of erythropoietin, megakaryocyte colony-stimulating factor, and T-cell-derived factors on human megakaryocyte colony formation: evidence for T-cell-mediated and T-cell-independent stem cell proliferation. Exp Hematol 1987; 15: 845-53.

[126]   Shingo T, Sorokan ST, Shimazaki T, Weiss S. Erythropoietin regulates the *in vitro* and *in vivo* production of neuronal progenitors by mammalian forebrain neural stem cells. J Neurosci 2001; 21: 9733-43.

[127]   Wada T, Haigh JJ, Ema M, *et al.* Vascular endothelial growth factor directly inhibits primitive neural stem cell survival but promotes definitive neural stem cell survival. J Neurosci 2006; 26: 6803-12.

[128]   Brazel CY, Nuñez JL, Yang Z, Levison SW. Glutamate enhances survival and proliferation of neural progenitors derived from the subventricular zone. Neuroscience 2005; 131: 55-65.

[129]   Rolls A, Shechter R, London A, *et al.* Toll-like receptors modulate adult hippocampal neurogenesis. Nat Cell Biol 2007; 9: 1081-8.

[130]   Koo JW, Duman RS. IL-1beta is an essential mediator of the antineurogenic and anhedonic effects of stress. Proc Natl Acad Sci USA 2008; 105: 751-6.

[131]   Widera D, Holtkamp W, Entschladen F, *et al.* MCP-1 induces migration of adult neural stem cells. Eur J Cell Biol 2004; 83: 381-7.

[132]   Sironi L, Banfi C, Brioschi M, *et al.* Activation of NF-κB and ERK1/2 after permanent focal ischemia is abolished by simvastatin treatment. Neurobiol Dis 2006; 22: 445-51.

[133]   Ninkovic J, Mori T, Götz M. Distinct modes of neuron addition in adult mouse neurogenesis. J Neurosci 2007; 27: 10906-11.

[134]   Mattson MP, Camandola S. NF-kappaB in neuronal plasticity and neurodegenerative disorders. J Clin Invest 2001; 107: 247-54.

[135]   Wang S, Scott BW, Wojtowicz JM. Heterogenous properties of dentate granule neurons in the adult rat. J Neurobiol 2000; 42: 248-57.

[136]   Schmidt-Hieber C, Jonas P, Bischofberger J. Enhanced synaptic plasticity in newly generated granule cells of the adult hippocampus. Nature 2004; 429: 184-7.

[137]   Song H, Kempermann G, Overstreet Wadiche L, Zhao C, Schinder AF, Bischofberger J. New neurons in the adult mammalian brain: synaptogenesis and functional integration. J Neurosci 2005; 25: 10366-8.

[138]   Barco A, Alarcon JM, Kandel ER. Expression of constitutively active CREB protein facilitates the late phase of long-term potentiation by enhancing synaptic capture. Cell 2002; 108: 689-03.

[139]   Pittenger C, Huang YY, Paletzki RF, *et al.* Reversible inhibition of CREB/ATF transcription factors in region CA1 of the dorsal hippocampus disrupts hippocampus-dependent spatial memory. Neuron 2002; 34: 447-62.

[140]   Duffy SN, Nguyen PV. Postsynaptic application of a peptide inhibitor of cAMP-dependent protein kinase blocks expression of long-lasting synaptic potentiation in hippocampal neurons. J Neurosci 2003; 23: 1142-50.

[141]   Jia Z, Lu YM, Agopyan N, Roder J. Gene targeting reveals a role for the glutamate receptors mGluR5 and GluR2 in learning and memory. Physiol Behav 2001; 73: 793-02.

[142]   O'Riordan KJ, Huang IC, Pizzi M, *et al.* Regulation of nuclear factor kappaB in the hippocampus by group I metabotropic glutamate receptors. J Neurosci 2006; 26: 4870-9.

[143]   Di Giorgi Gerevini VD, Caruso A, *et al.* The mGlu5 metabotropic glutamate receptor is expressed in zones of active neurogenesis of the embryonic and postnatal brain. Brain Res Dev Brain Res 2004; 150: 17-22.

[144]   De Smaele E, Zazzeroni F, Papa S, *et al.* Induction of gadd45beta by NF-kappaB downregulates pro-apoptotic JNK signalling. Nature 2001; 414: 308-13.

[145]   Ma DK, Jang MH, Guo JU, *et al.* Neuronal activity-induced Gadd45b promotes epigenetic DNA demethylation and adult neurogenesis. Science 2009; 323: 1074-7.

[146]   Ma QH, Futagawa T, Yang WL, *et al.* A TAG1-APP signalling pathway through Fe65 negatively modulates neurogenesis. Nat Cell Biol 2008; 10: 283-94.

[147]   Cao X, Südhof TC. Dissection of amyloid-beta precursor protein-dependent transcriptional transactivation. J Biol Chem 2004; 279: 24601-11.

[148]   von Rotz RC, Kohli BM, Bosset J, *et al.* The APP intracellular domain forms nuclear multiprotein complexes and regulates the transcription of its own precursor. J Cell Sci 2004; 117: 4435-48.

[149]   Zhao Q, Lee FS. The transcriptional activity of the APP intracellular domain-Fe65 complex is inhibited by activation of the NF-kappaB pathway. Biochemistry 2003; 42: 3627-34.

[150]   Baek SH, Ohgi KA, Rose DW, Koo EH, Glass CK, Rosenfeld MG. Exchange of N-CoR corepressor and Tip60 coactivator complexes links gene expression by NF-kappaB and beta-amyloid precursor protein. Cell 2002; 110: 55-67.

[151]   Caillé I, Allinquant B, Dupont E, *et al.* Soluble form of amyloid precursor protein regulates proliferation of progenitors in the adult subventricular zone. Development 2004; 131: 2173-81.

[152]   Grilli M, Goffi F, Memo M, Spano PF. Interleukin-1beta and glutamate activate the NF-kappaB/Rel binding site from the regulatory region of the amyloid precursor protein gene in primary neuronal cultures. J Biol Chem 1996; 27: 15002-7.

[153]   Mattson MP, Meffert MK. Roles for NF-kappaB in nerve cell survival, plasticity, and disease. Cell Death Diff 2006; 13: 852-60.

[154]   Toliver-Kinsky T, Papaconstantinou J, Perez-Polo JR. Age-associated alterations in hippocampal and basal forebrain nuclear factor kappa B activity. J Neurosci Res 1997; 48: 580-7.

[155]   Greenberg DA. Neurogenesis and stroke. CNS Neurol Disord Drug Targets 2007; 6: 321-5.

CHAPTER 6

NF-κB Transcription Factor: A Model for the Study of Transcription Regulation in Memory Consolidation, Reconsolidation and Extinction

Arturo Romano[*]

Laboratorio de Neurobiología de la Memoria, Departamento de Fisiología, Biología Molecular y Celular. Facultad de Ciencias Exactas y Naturales, Universidad de Buenos Aires, IFIBYNE-CONICET, Ciudad Universitaria, Pab. II, 2do Piso (1428) Buenos Aires. Argentina

Abstract: Some decades ago it was postulated that gene expression regulation is required for long-term memory storage. In the last years, important progress has been made towards the characterization of these mechanisms of transcription. Among them, transcription factors play a key role. This chapter describes the characteristics and the role of one of these transcription factors, NF-κB, in the different phases of memory formation and processing. In more than a decade of research since the first data on the role of NF-κB in neuronal plasticity and memory, a growing body of evidence supports that this transcription factor is involved not only in the formation of the initial long-term memory traces but in the re-stabilization after memory reactivation induced by retrieval. Additionally, the role of NF-κB in the formation of memory extinction is now under study. Extinction entails a temporary inhibition of memory expression and entails a new memory process. Recent data support the inhibition of NF-κB in the formation of memory extinction. Here I propose the use of this transcription factor, together with other neuroplasticity-associated molecular mechanisms, as important tools for understanding the dynamics of the different phases of information processing as well as the determination of brain areas involved in such processes.

Keywords: Long-term memory, consolidation, reconsolidation, extinction, gene expression, transcription factors, NF-kB, IKK, kB decoy, sulfasalazine.

INTRODUCTION

Since the pioneering studies of memory deficits in patients with brain damage carried out by Ribot, the concept has raised that memory can be disrupted, and eventually lost, by means of disrupting agents. The first disruptors to be studied were brain trauma and convulsive electroshock. The use of experimental animal models allowed a more precise characterization of time windows and conditions for memory disruption, defining more accurately the concept of consolidation. Consolidation was defined as a temporary limited process required to convert memory from an initial, labile and short-term lasting stage to a stable, long-term lasting one [1]. During the second half of the last century an important step could be made towards the mechanistic characterization of consolidation, when antibiotics that inhibit mRNA translation to protein were identify as amnesic agents during the consolidation phase. Accordingly, consolidation was defined as a memory phase in which new protein must be synthesized [2]. Different interpretations were made from this fact. Why is then a critical period for protein synthesis? In 1986, a suggestive paper proposed an answer to this question in terms similar to those for cell differentiation processes [3]. In fact, there are some critical periods during development in which *de novo* protein synthesis is required. This is so because extracellular signals induce immediate-early gene (IEG) expression, whose messengers must be translated into protein in such critical periods for differentiation to take place. In a similar way, it was proposed that memory consolidation requires IEGs expression which must be translated for long-term memory formation (Fig. **1**).

*Address correspondence to Arturo Romano: Laboratorio de Neurobiología de la Memoria, Departamento de Fisiología, Biología Molecular y Celular. Facultad de Ciencias Exactas y Naturales, Universidad de Buenos Aires, IFIBYNE-CONICET, Ciudad Universitaria, Pab. II, 2do Piso (1428) Buenos Aires. Argentina; E-mail: aromano@fbmc.fcen.uba.ar

However, *de novo* protein translation does not necessarily involved *de novo* gene expression. For instance, long-term potentiation (LTP), a model for neural plasticity in glutamatergic synapses widely employed to explain – at least in part – cellular and molecular features of memory, shows a translation-depending but not transcription-depending phase. In fact, late LTP can be divided into an initial translation-depending phase and a later, transcription-depending period which allows for *in vivo* potentiation lasting 1 day or more [4]. This fact suggests that translation of mRNA synthesized previously to synaptic activity is enough for the initial phase of late LTP and gene expression is only required for more persistent plasticity. Local synaptic translation is proposed as an important mechanism in neural plasticity and can be carried out independently from transcription by translation of dormant mRNA [5].

Recently, the key role of gene expression and protein synthesis were challenged [6, 7]. Protein synthesis would not be instructive for memory formation but only for replenishment of the used proteins during neuronal activity. In that proposal, only post-translational modifications of synaptic proteins would be instructive for memory. However, covalent post-translational modifications do not completely account for the maintenance of long-term memory, and transcription of specific genes, translation in specific time and place and post-translational modification, would be all necessary for long-term plasticity. At our point of view, transcription regulation is strictly necessary not only for long-term memory formation, but also for memory reprocessing after retrieval.

There is wide evidence for transcription in consolidation since the pioneering work on brightness discrimination in rats [8. 9], consolidation in the chicken avoidance task [10], long-term facilitation in the mollusk *Aplysia* [3, 11] and in several more recent reports ([12], for a revision).

At present, an important body of experimental data is available involving a group of five key transcription factors (TF) in neural plasticity and memory. To be considered as such, we have taken into account the following criteria:

1) Evidence of correlation, in which the involvement of these TF in memory formation is clearly established. Such evidence is obtained using adequate controls in order to differentiate specific activation due to intrinsic memory process than unspecific activation due to indirect processes, such as stress, motor activity, sensorial stimulation, etc.

2) Evidence of the requirement of these TF by means of interventive treatments, in which inhibition or activation of the TF or the upstream pathways are used. After intervention in precise time windows of the phase of memory under study, the effect in behavior is evaluated in a posterior test. The inhibition of the TF should impair the phase of memory, while the activation should improve memory. However, as is discussed in the section D, inhibition of the TF could improve the memory phase, depending on the role of that mechanism of transcription in that phase.

3) Determination of the previous criteria in different memory model, particularly in phylogenetically distant species.

Among the TF that fulfill the above mentioned criteria, two of them, cAMP response element binding protein (CREB) and nuclear factor κB (NF–κB), show relatively high basal expression, and that characteristic warrant the presence of the proteins before the neuronal activity involved in the acquisition of the information. This fact implies that the TF can be activated promptly by synaptic activity and extracellular signals by protein-protein interaction or covalent modifications, usually phosphorylations. In this way, the presence of the TF in regulatory regions of different genes and the activation of transcription can be rapidly achieved. Although some genes that codify for CREB and NF-κB can be induced by extracellular signals, here we will consider these TF as constitutive for its high basal expression (Fig **1**).

The other 3 TF, C/EBP, ZIF268 (also known as ERG-1) and AP-1, are Immediately-Early Genes (IEG). The basal expression of their genes is very low (at the limit of detection), and required induction in order to reach the level of protein necessary for the action of the TF in the nucleus (Fig. **1**). There are evidence in memory processes that C/EBP is regulated, at least in part, by CREB [13], while NF-κB is involved in the regulation of ZIF268 [14].

In the initial search for the role of TFs in memory consolidation, the analysis was particularly focused on CREB family of constitutive TFs. Work performed in rodents, Aplysia and Drosophila, supports that CREB is an evolutionary conserved mechanism involved in neural plasticity associated with memory formation (*e.g.*, [15-17]). However, regulation of gene expression is a multifactorial process that entails the involvement of a defined set of TFs and coactivators [18]. A comprehensive view of gene expression regulation in the nervous system, particularly regarding memory formation, is that parallel pathways of gene induction are recruited [19] as suggested by microarray studies in animals after training [20, 21, 22]. The present chapter will focused on NF-κB TF family, considered as a component of a integrated system.

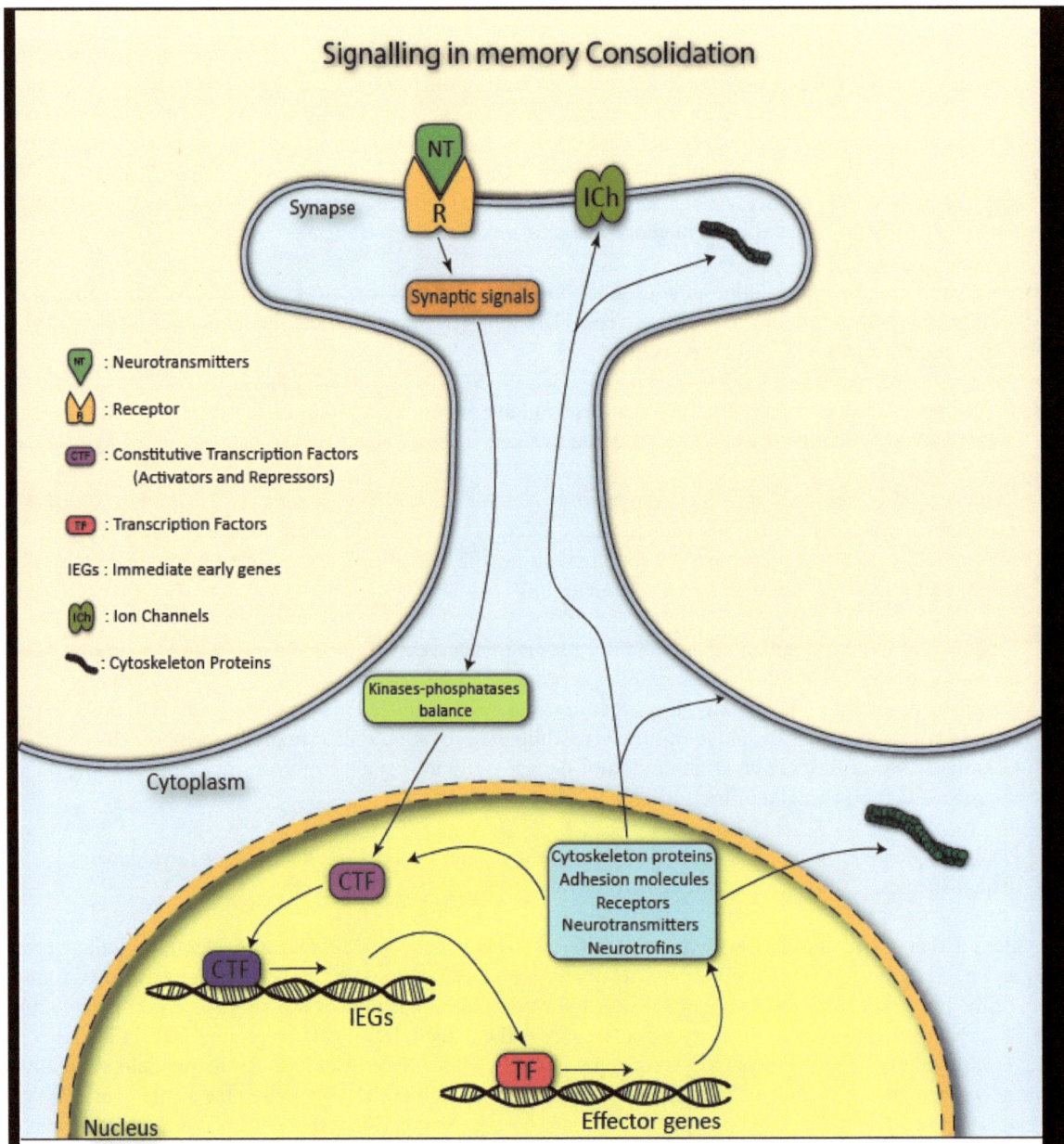

Fig. (1). Schematic representation of the molecular pathways involved in memory formation. Synaptic activity during learning induces second messengers that change the balance between protein kinases and protein phosphatases activity, usually to kinases activation. When synaptic activity is enough to overcome a threshold, the signals reach to the nucleus, either by TF or protein kinases translocation. Constitutive TF are rapidly activated by covalent modification, normally by phosphorylation. IEG are then induced. Some of them are TF, and together with constitutive TF regulate late gene expression. Late genes products are protein involved in the modification and maintenance of synaptic efficacy.

CREB and NF-κB share the common feature of being fast responders to extracellular signals and of being involved in the regulation of immediate-early genes required for memory formation. They also share the interaction with histone acetyl transferases (HAT), such as CREB binding protein CBP and p300 [23, 24]. Histone acetylation is a post-translational modification of chromatin being one of the epigenetic mechanisms for long-lasting regulation of gene expression. The initial step in gene-transcription activation involves recognition of the promoter and enhancer regions by sequence-specific binding of the TFs to DNA. These TFs can also recruit chromatin remodeling factors to gene regulatory regions. Chromatin-remodeling factors such as acetyltransferases and histone phosphatases catalyze the mobilization of nucleosomes, necessary for transcription initiation. In mature neurons, some genes may be available for transcription, while others will only be transcribed when stimulation and/or modulation are sufficient for major gene regulation. A example out of the area of neurobiology is the two waves of NF-κB transcriptional regulation that were found after LPS stimulation in macrophage cells. One of them is a fast recruitment of NF-κB to immediately-accessible genes, while the other corresponds to a late recruitment of NF-κB to a set of promoters that require histone acetylation to make them accessible [25]. In the last years, a growing body of evidence supports the involvement of epigenetic mechanisms in long-term memory [26, 27].

The NF-κB System of Gene Expression Regulation

In contrast to the transcription factors belonging to the immediate-early gene group that require translation, NF-κB is normally preexisting in the cytoplasm in an inactive form. In this state, the TF is bound to its inhibitor, IκB, which occludes the nuclear localization sequence (NLS), a sequence that regulates nuclear translocation. In response to specific stimuli, NF-κB translocates to the nucleus as a crucial step in the regulation of its transcriptional activity [28]. In mammals, the NF-κB family of TF comprises five members that share the Rel homology domain (RHD) and that form homo- and hetero-dimers. The heterodimer p65 (Rel A) / p50 is the most common form of the TF (Fig 2).

Rel/NF-kB transcription factor was first characterized in lymphocytes [29] but is an ubiquitously expressed protein with high level of expression in the nervous system. All NF-κB proteins share the Rel homology domain (RHD). In mammals, the family comprises NF-κB 1, NF-κB 2, p50 (a product of the NF-kB1 gene), p52 (a product of the NF-kB2 gene), Rel A (p65), cRel/vRel, and Rel B. Members of the Rel/NF-κB family bind DNA as homo or heterodimers, recognizing the consensus decameric sequence 5′-GGGPuNNPyPyCC-3′, designated as κB site (Fig 2). Different dimers bind with diverse affinity to the distinct κB sites. In addition, gene transcription can be either activated or repressed depending on the dimer composition. For example, heterodimers composed by Rel A and p50 (or p52) are activators, while p50 or p52 homodimers are repressors. The presence of Rel/NF-kB members was found in different species such as human, rat, mouse, chick, *Xenopus*, *Drosophila*, *Anopheles*, honeybee, crab, oyster and *Aplysia*, indicating that this TF family is evolutionarily conserved.

NF-κB is widely distributed in the CNS and it is activated by signals involved in neural plasticity such as glutamate through NMDA receptors [30], depolarization, transient increment of intracellular Ca^{2+} (reviewed in [31]), neuropetides such as angiotensin II [32], neural cell adhesion molecule (N-CAM) and cytokines (TNFα and IL-1) [33, 34]. Several genes whose protein products are involved in memory storage contain NF-κB recognition sites in the promoter regions, and in many of them it was demonstrated that NF-κB is in fact involved in its transcription regulation. Among them are the inducible nitric oxide synthase (iNOS), the neural cell adhesion molecule (N-CAM), angiotensinogen, cytokines, opiod peptides (reviewed in [31]), the AMPA glutamate receptor [35] and brain derived neurotrophic factor (BDNF) [36] (Fig. 2).

The Role of NF-κB in the Synapse to Nucleus Communication

As mentioned in the previous section, NF-κB has an interesting feature, i.e., in resting conditions this TF is retained in an inactive form out the nucleus by binding to IkB inhibitory protein. This peculiarity is particularly relevant in neurons, in which the distance between axonic and dendritic processes and cell soma is important. NF-κB is localized not only in the soma cytoplasm, but also at the synaptic compartment [37-39]. This finding opened to the question of what is the role of this TF pool located so far to the normal site of action in the nucleus. One hypothesis is that NF-κB plays a role in the communication between synapse activity and DNA transcription. The presence of

NF-кB in synapses would allow its local activation and then, by retrograde transport, its translocation to the nucleus. This fact should confer to this TF a dual function, as synaptic activity detector and as transcriptional regulator.

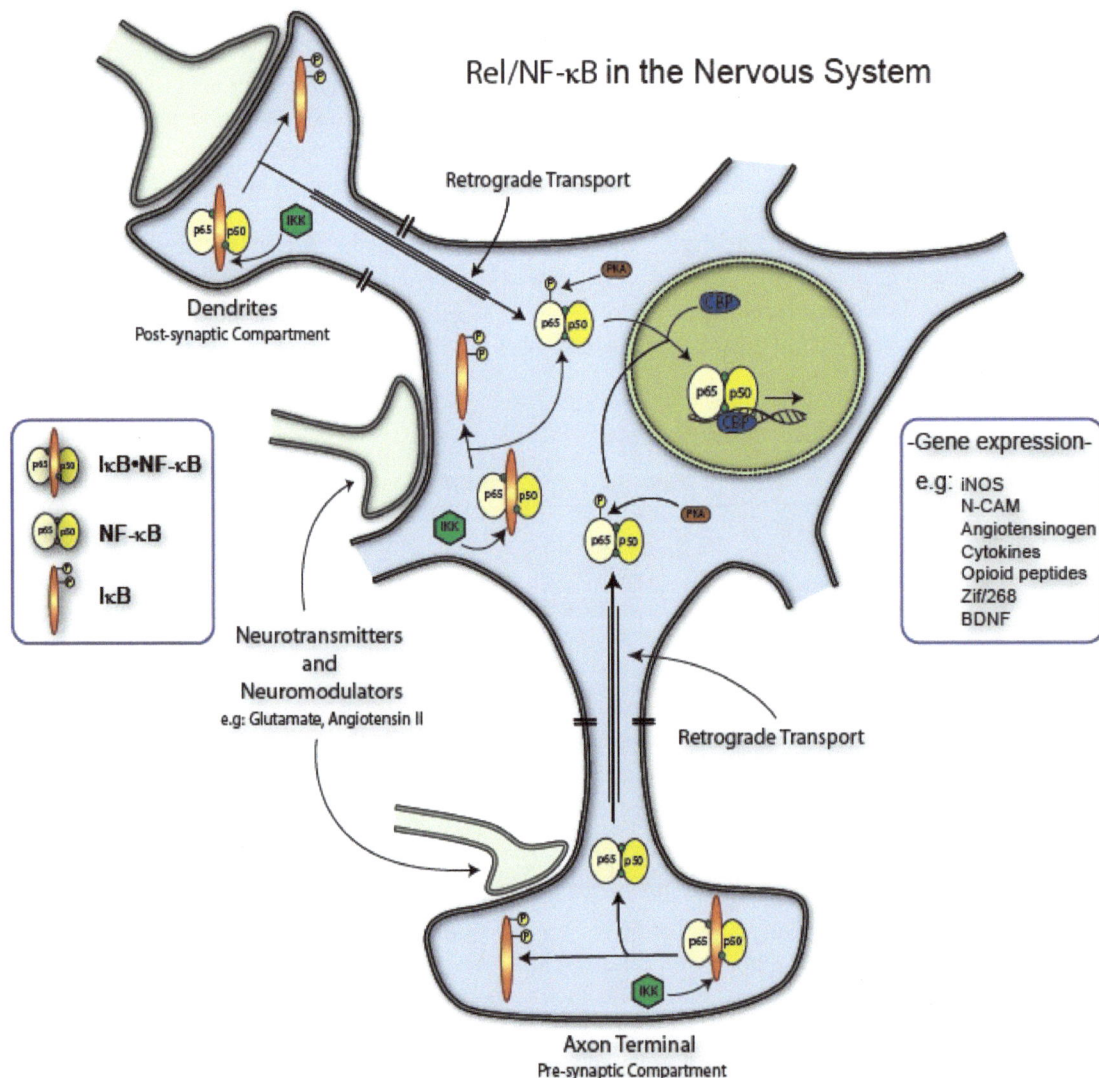

Fig. (2). Summary drawing of NF-кB pathway activation by synaptic activity and the regulation of transcription during memory formation. NF-кB, represented here as p65/p50 heterodimer (the common form in mammals), is activated by neurotransmitters or neuromodulators that induce the phosphorylation of inhibitor protein IкB and its subsequent degradation. Depending of the site of activation, the signals will arrive to the nucleus at different times to confer a fast or a delayed transcriptional response after retrograde transport. After IкB releasing, NF-кB undergoes covalent modifications such as phosphorylation by PKA that regulate nuclear translocation and/or transcriptional activity. PKA phosphorylation of p65 allows CBP binding and lysine residues acetylation, which modulates NF-кB transcriptional activity.

In eukaryotic cells, protein transport into the nucleus is mediated by short amino acid sequences that are referred to as nuclear localization signals (NLS). Classical NLS-dependent nuclear transport relies on transport factors that belong to the importin family of proteins. Importin α binds NLS of the cargo protein directly, and its affinity for NLS is increased by interaction with importin β. Importins are found throughout axons and dendrites and play a critical role in transporting synaptically generated signals towards the nucleus. The trimeric import complex containing the NLS-bearing cargo, importin α and importin β was found to traffic retrogradely due to an interaction of importin α with the motor protein dynein [40].

The importance of retrograde transport in synaptic plasticity was stressed by a work of Martin's laboratory [41] in which they investigated whether the classical active nuclear import pathway mediates intracellular retrograde signal transport in *Aplysia* sensory neurons and rodent hippocampal neurons. In both types of synapses, they found that stimuli known to produce long-lasting but not short-lasting plasticity triggered importin nuclear translocation. NF-kB members have the NLS in the RHD region and, in particular, p65 is associated with components of the dynein complex *in vivo* [42].

The first evidence supporting the synapse-to-nucleus function of NF-kB came from *in vitro* cellular studies. Kaltschmidt's group obtained the first results showing that glutamate or kainate stimulation and depolarization in hippocampal primary neurons cultures provoked a clear redistribution of p65 fused with the fluorescent EGFP protein from neurites to nucleus [43]. In another work from the Baltimore's group [30], glutamate stimulation increased NF-κB activity, and NMDA glutamate receptor and L-type Ca^{2+} signal inhibitors reduced basal activity. In this work only the p65/p50 dimer was detected in synaptosomes. A p65-GFP fusion protein was detected in cytoplasm and dendrites, including spine-like structures. As in the work of Kaltschmidt and colleagues, glutamate stimulation caused a large reduction of fluorescence in dendrites along with nuclear fluorescence increment 20-30 min after stimulation. Further studies with photobleaching led to the conclusion that p65-GFP from distal processes reached the nucleus 30-45 min after stimulation. These data suggest that, in fact, NF-κB is recruited from synapse to nucleus after gutamatergic activation, and that the process is dependent on NMDA receptor activation.

NF-κB in Neural Plasticity

The study of neural plasticity in laboratory models such as long-term potentiation (LTP) in rodents and long-term facilitation in sensory-motor neuron synapse have provided important insight for the understanding of molecular processes involved in memory and, in particular, for the study of the gene expression regulation. Two phases have been described for late LTP: LTP1 is translation- but not transcription-dependent, LTP2, is dependent of transcription. LTP2 is the long-lasting phase of LTP that is observed *in vivo* for one or more day [44]. Gene transcription is well characterized in LTP. Contrarily, there are few studies on the involvement of gene expression and transcription factors in long-term depression (LTD), the neural plasticity counterpart mechanism, which has been studied mainly in hippocampal excitatory synapses [45, 46].

The first experimental results linking NF-κB with synaptic plasticity were obtained in relation to the hippocampal long-term potentiation (LTP) using ipsilateral tetanus of the rat perforant path *in vivo* [39]. In that study, increased levels of p50 and p65 mRNA, relative to the contralateral side, were found in both LTP and low frequency controls 60 min after stimulation. This indicates that after neuronal activity the expression of both genes is induced but p65 mRNA in granule cells was significantly greater in the LTP animals relative to the low frequency controls. Albensi and Matson found that the stimulation of hippocampal slices in Schaffer collateral axons at low frequency (1 Hz) induced long-term depression that was prevented by infusion of the NF-κB inhibitor κB decoy DNA, whereas the same treatment significantly reduced the amplitude of LTP induced by HFS [47]. In amygdala slices, NF-κB and IκB kinase (IKK) activity increased after tetanus, and kB decoy DNA infusion impaired LTP induction [48]. In another study, the activation of NF-κB found in mouse hippocampus 15 min after tetanization of the perforant path was compared with control animals by two independent methods, electrophoretic mobility shift assay (EMSA) with nuclear extracts from total hippocampus and immunohistochemistry with an activated-specific p65 antibody that recognizes the NLS region [49]. Taken together, these studies support that NF-κB is one of the TFs involved in synaptic plasticity.

NF-κB IN MEMORY CONSOLIDATION

Although the use of transcription inhibitors as memory disruptors in consolidation was less extensive than the use of protein synthesis inhibitors, the memory impairment induced by drugs as actinomicin D applied shorter after information acquisition has suggested that gene transcription is necessary for the formation of long-term memory. The difficulties in using transcriptional inhibitors are their toxicity and unspecific effects. However, many studies in diverse species have revealed that mRNA synthesis is necessary during 1-2 hours after training. Indeed, a second phase of transcription was described to be necessary in rats and crabs [50], Merlo and Freudenthal, personal communication) 3 to 6 h after training.

The inhibition of specific transcription factors provides more precise information about the particular mechanisms that are involved in memory consolidation. For instance, the use of decoy DNA oligonucleotides that contain a consensus sequence allows for specific transcription factors inhibition. This strategy was initially used to study the role of CREB and C/EBP in Aplysia neuroplasticity [51, 13]. After that, other strategies were used for specific TFs inhibition in memory and neuroplasticity, such as antisense oligonucleotides [52, 53], inducible transgenes and knock-outs [54], siRNA and inhibition of protein kinases that are involved specifically in TF activation [55].

NF-κB in Invertebrates' Memory Processes

The first evidence that links NF-κB family function with memory formation was obtained in invertebrates, in studies on the long-term memory model of the crab *Chasmagnathus*. In this model, the repeated presentation of a visual danger stimulus (an opaque figure passing over the animal) induced the replacement of the initial escape response by a freezing response [56, 57]. Fifteen or more spaced presentation of the figure induce an association between the iterated stimulus (unconditioned stimulus, US) and contextual features (conditioned stimulus, CS) [58]. A long-term memory is then formed, termed context-signal memory (CSM), which lasts at least for a week and is blocked by protein and mRNA synthesis inhibitors [59, 60]. With massed training, in contrast, habituation (a non-associative form of learning) was induced [61, 62].

Results obtained with this invertebrate learning model have evidenced the participation of NF-κB in CSM formation [63, 64]. CSM formation correlated with NF-κB activation in the nucleus of the crab central brain. On the contrary, NF-κB is not activated after massed training. Two phases of NF-κB activation were found in crabs that received spaced training. The first phase was found immediately after training and decayed to basal level 3 h after training. The second phase was found 6 h after training, waning to basal level in 12 - 24 h [64]. Thus, in this model a fast and transient activation was followed by a prolonged phase of activity (Fig. **1B**). This finding is in coincidence with the two phases of mRNA synthesis in rat avoidance memory already described [50]. A similar activation profile was found in other systems like tumor necrosis factor α (TNFα) NF-κB activation in promyelocyte cultures, with peaks at 4 min and 3 h, both dependent on different pathways. The first transient phase of activity is thought to be caused by rapid IκB phosphorylation and proteolysis in response to transducing signals. The second long-lasting phase is considered to be dependent on protein synthesis and NF-κB genes induction may be regulated by its own protein products in an autoregulatory loop [65]. Unpublished results obtained in *Chasmagnathus* by administering the translation inhibitor cycloheximide showed that the second phase of NF-κB activation can be attenuated by protein synthesis inhibition (Freudenthal, Merlo & Romano, unpublished results), supporting that the second phase depends on a positive feed-back loop acting on NF-κB genes transcription. Coincidently, microarray analysis of gene expression during consolidation of long-term fear-contextual memory in hippocampus revealed an initial burst of gene upregulation (1h after training) that was followed by a period of gene repression (2-4 h) and a subsequent second wave of gene expression. NF-κB consensus sequence was found among the regulatory elements shared with most of the memory-related genes [22]. In other work, the presence of NF-κB recognition element in promoters of some genes was studied using chromatin immunoprecipitation assay (ChIP), a useful method to study the recruitment of TFs and other regulatory proteins in target promoters [25]. After LPS stimulation in macrophage cells, two waves of NF-κB transcriptional regulation were found. One of them is caused by a fast recruitment of NF-κB to immediately-accessible genes and the other one corresponds to a late recruitment of NF-κB to a set of promoters that require histone acetylation to make them accessible. Thus, the two phases of nuclear NF-κB activity found after learning may regulate different sets of genes, the first of them initially required for a fast response and the second phase, induced by the positive feed-back loop, necessary for more long-lasting response.

In the crab model, an NF-κB inhibitor has been shown to impair memory consolidation. Sulfasalazine, an inhibitor of IκB kinase (IKK), induced amnesia when administered during the two periods in which NF-κB was active, but not after such activation periods [55]. Sulfasalazine also has anti-inflammatory effects and its action was initially attributed to COX inhibition. However, the purported effect on prostaglandin levels was controversial and, in general, sulfasalazine seems to be a weak or very weak inhibitor of COX [66], but proved to be a potent and specific inhibitor of NF-κB [67] due to interference with the ATP binding site of IKK [68]. Thus, the anti-inflammatory effect of sulfasalazine is attributed to NF-κB inhibition. Merlo and coworkers [55] found that the administration of a COX inhibitor, indomethacin, did not affect *Chasmagnathus* CSM. The last result suggests that sulfasalazine effect on memory is due to inhibition of NF-κB.

NF-κB in Mammalian Memory Models

A growing body of evidence support that NF-κB is a significant component of the molecular mechanism of memory in mammals. The first evidence was obtained by Gean and colleagues studying fear conditioning in rat. In that task, conditioned animals received 10 light-foot shock pairings and controls received the same number of light and foot shock presentation but in a pseudorandom fashion. In a testing session, the fear-potentiated startle to a white noise is evaluated in the presence of the conditioned stimulus. Using this model, Yeh and collaborators [48] found by EMSA an increment of NF-κB DNA-binding activity 2 to 6h after training in amygdala but not in hippocampus or cerebellum, two structures not involved in the neural representation of this learning. IKK activation was also found at 10 and 30 min after training, decreasing to basal level at 90 min. A double intra-amygdala administration of κB decoy 2 and 24 h before training significantly reduced the response at testing, supporting that NF-κB activation in the amygdala is required for memory storage in this fear conditioning paradigm. Further studies in this model revealed that fear conditioning induced an association between NF-κB and CREB binding protein (CBP), resulting in an increment of acetylated p65 due to the acetyl-transferase properties of CBP. Administration of general deacetylase inhibitors prolonged the acetylation state of p65 in the amygdala and facilitated long-term memory [69]. Previous to these findings, Zhong [70] and colleagues had described the association between CBP and NF-κB, recruiting the acetyl-transferase to promoters containing κB sites.

Further evidence for the role of Rel/NF-kB was obtained in hippocampus in an inhibitory avoidance task in mice. In this task, animals are placed on a platform in front of the entrance to a dark compartment. Trained animals receive a mild foot-shock in the compartment, while controls do not receive it [71]. At testing, trained animals typically show high latencies to step-through or do not come into the compartment. Conversely, control animals show very low latencies to entering. Sulfasalazine post-training administration induced a dose dependent retention deficit when tested 48 h later [72]. Conversely, delayed injections of sulfasalazine at 3 or 24 h post-training or the COX inhibitor indomethacine did not affect retention. Thus, as previously found in the crab model, sulfasalazine induced retention deficit acting specifically on the NF-κB pathway. Then, we employed κB decoy i.c.v. administration 2 h pre-training, in order to study the effect of direct NF-κB inhibition. In agreement with the previous evidence, memory impairment was found. Conversely, injection of κB decoy with one base mutation did not affect long-term memory. κB decoy actually inhibited NF-κB in hippocampus 2 h after injection, but no inhibition was found with mut-κB decoy administration. A temporal course of hippocampal NF-κB activity after training showed initially an NF-κB inhibition phase 15 min after training in shocked and unshocked groups when compared with naïve group. The inhibition was followed by activation 45 min after training in both groups, returning to basal levels 2 and 4 h after training. These results would support that the NF-κB activation in hippocampus is required for storage of contextual features that constitute the conditioned stimulus representation. Further experiments in which both experimental device and foot-shock were presented for 7 days but in a non contingent manner (foot-shock was presented in a different place and mice were removed immediately after to avoid shock-context association) indicated that NF-κB was no longer activated. However, in the following day foot-shock was presented in the experimental device, and in such conditions NF-κB was activated in hippocampus. This result supports the hypothesis that NF-κB activation in the hippocampus participates not only in contextual features storage, but also in the plasticity mechanism required for long-term storage of the CS-US association [73].

KO Mice for NF-κB Components

The use of gene targeting strategy in mice provided further evidence on the role of NF-κB in learning and memory. Mice with a null mutation of p50 showed acquisition deficits in an active avoidance task in which the animal must jump to a platform to avoid the shock [74]. Although this result would suggest that p50 is required for normal learning, no differences were found between knock-out and wild type mice in a test for long-term memory performed 24 h after training, suggesting that p50 is not required in memory formation but in acquisition. These null mutants are viable and fertile and do not have gross developmental and histological defects, but do in fact show immune deficits. A more recent study characterized this mutant at the behavioral level, showing decreased anxiety-like responses, elevated exploratory behavior and a reduced tendency to develop dominant-subordinate behaviors [75]. Rather than memory deficits, such relevant behavioral differences between mutants and wild type animals could then account for the differences in learning performance found in the previously mentioned work. In addition, the lack of p50 in knock-out animals prevent a clear prediction about its effect on κB dependent transcription, as it is a component of both an active transactivator (p65/p50 heterodimer) and a repressor (p50/p50 homodimer).

Baltimore and his group have found that deletion of the p65 subunit is lethal during embryogenesis [76], but the knockout can be rescued by concurrent deletion of type 1 TNF receptor [77]. The brain of p65/TNFR1 double-knockout mice appeared normal under general examination, and no obvious behavioral abnormalities were described. In a recent work, Meffert and co-workers analyzed spatial memory performance of this mutant using the spatial and cued versions of the radial arm maze. In the spatial version of the task, all of the 8 arms have positive reinforcement (food pellets in the external end of each arm) and extra-maze spatial cues are available. Using the cues for spatial orientation animals must remember and avoid the already visited arms in order to optimize food rewards. In the cued version of the task, no extra-maze clues are available and only the lit arms have food pellets. The animals must associate light with food and spatial orientation is not required. During the first sessions, p65-/-/TNFR1-/- mutants made more errors than p65+/+/TNFR1-/- siblings, but reached in the last sessions the same low level of errors than controls. On the contrary, no performance deficit was found in the cued version of the task. Spatial version of this paradigm depends on hippocampus, and thus, this result suggest that p65 is required in hippocampus for proper processing and storage of spatial information, whereas is not necessary in other areas such as striatum that are involved in non-spatial learning [30].

Levenson and colleagues [22] analyzed a c-Rel knock-out mouse (c-Rel-/-) in two versions of a fear conditioning paradigm. In the cued version animals are lodged in an experimental cage and a white noise is paired with foot-shock. During testing, freezing response to the noise and amygdala function are evaluated. No differences between c-Rel knock-out and wild type were found in this study, indicating that amygdala-dependent long-term memory is unaffected. This conclusion differs from that obtained by Gean and colleagues using another form of cued fear conditioning, in which it was found that an activation of NF-κB in amygdala is necessary for memory formation [48]. However, c-Rel protein seems to be necessary in the contextual version of the paradigm in which a novel context is paired with foot-shock. For this version, the hippocampal function is required and a freezing response is induced when animals are re-exposed to the training context. In this task c-Rel knock-out showed impaired contextual memory.

More recently, a forebrain NF-κB deficient mouse was used to test spatial memory deficits (78). These authors generated a double transgenic mouse with the expression of tetracycline transactivator (tTA) under the CamKIIα promoter, which confers postnatal expression restricted to forebrain. The second transgene includes a TetO promoter, regulated by tTA, which allows expression of a non-phosphorylatable IκBα form (IκB super-repressor). This double transgene renders the animal NF-κB deficient in forebrain neurons. The treatment with doxycyline, a permeant analog of tetracycline, rescues the NF-κB deficiency. The performance of this double transgenic mouse without doxycycline (NF-κB deficient) was tested in Morris maze for spatial memory. The animals improved the latency to swim to the hidden platform during the 9 daily sessions of training, but showed higher latencies than IκBα single transgenic controls. Similar differences between these two groups were found in a transfer test without the platform, in which the time spent searching in the right quadrant was determined. However, no differences were found between NF-κB deficient animals and doxycycline treated animals (NF-κB rescued) due to an unspecific influence of the drug on memory formation. Activity dependent synaptic plasticity was reduced in Schaffer CA3-CA1 hippocampal pathway after induction of LTP and LTD. An important finding of this work is that several genes are downregulated in NF-κB deficient animals, such as junD immediate-early gene, neuronal cell adhesion molecule (N-CAM) gene and PKA catalytic α subunit gene.

Therefore, with the exception of the work of Gean and colleagues on rat amygadala, the available bibliography points to an important role of NF-κB in activity dependent synaptic plasticity of the hippocampal formation. Such a role is reflected in the participation of this TF in spatial and contextual memory tasks that depend on the hippocampal function, such as radial arm maze, Morris water maze, fear conditioning and inhibitory avoidance, whereas non-spatial or cued tasks that are not dependent on hippocampal function showed no impairment in NF-κB deficient animals. Additional evidence on the role of NF-κB in hippocampus is described in the "NF-κB role in reconsolidation" section.

Neurotransmiters and Neuromodulators that Activate NF-kB in Neuroplasticity and Memory

The finding that NF-κB is activated during memory consolidation and that its inhibition impairs memory led to the question of which neural pathways and neurotransmitters are involved in the activation of NF-κB. As mentioned previously, the involvement of the transcription factor in hippocampal-dependent memory tasks and in synaptic plasticity in hippocampus point to the activation of NF-κB by glutamate and, particularly, through the NMDA receptor [30, 23]. BDNF plays a crucial role in LTP and their effect is dependent in part on NF-κB activity [79]. Among other putative NF-κB-activating neurotransmitters and neuropeptides, angiotensin II (AII) has been associated with memory processes, and in *Chasmagnathus* it plays an important role in long-term memory formation. Pre-training administration facilitates memory formation and the AII receptor antagonist saralasin impairs long-term memory [80]. Furthermore, AII was proposed as a regulator of several physiologic functions in order to produce an orchestrated response to natural challenges such as water deprivation and increasing of water salinity. In fact, high water salinity and water shortage provoke an increase in the endogenous AII that correlates with memory facilitation [32, 81], and such facilitation is reversed by saralasin. Nuclear NF-κB in the crab brain is activated by ANGII, and this effect is reversed by saralasin. This activation is also found after water shortage, in coincidence with memory facilitation [32].

Downstream Effectors of NF-κB Associated with the Formation and Expression of Memory

The participation of NF-κB in the regulation of the expression of many genes has been analyzed during the last decade. Some of them were studied in other cells than neurons and the implication in the nervous system should be determined. However, an important number of genes whose protein products are involved in neural plasticity and memory are regulated in part by this transcription factor. Among them are protein kinases: PKC δ and PKA catalytic α subunit; neurotransmitters and neuropeptide receptors: NMDA receptor 1 and 2A, μ and δ opioid receptor and insulin-like growth factor receptor; immediate-early genes: Erg-1/Zif268, c-Fos, JunB and JunD; neuropeptides: angiotensinogen, proenkephalin and proopiomelanocortin; components of the proteasome: proteasome subunit LMP2; cytokines: interleukins, TNFα and β and their receptors; enzymes: inducible NO synthase (iNOS) and cell adhesion molecules: neural-cell adhesion molecule (N-CAM) and the brain-derived neurotrophic factor (BDNF). The expression of some of these genes was studied in the context of memory formation and/or expression [22, 78], and more studies are required to determine the specific role of NF-κB in the regulation of these genes during memory processes. In a comprehensive study, Lubin and colleagues [14] recently demonstrated the role of NF-kB in the control of BDNF expression. The synthesis and release of BDNF in hippocampus 12 h after training plays a key role in memory persistence [82, 83].

NF-κB IN MEMORY RECONSOLIDATION

In the classical postulation, memory is stored in its definitive long-term form once consolidated few hours after acquisition, becoming resistant to different disruption treatments. However, a body of evidence has challenged this assumption, supporting the view that memory can be reactivated by a reminder after retrieval, and that a new period of lability followed by a consolidation like-processes are engaged [84-88]. This phenomenon, known in literature as reconsolidation, is under intense research and it is a matter of debate whether its differences to the initial consolidation [89-92]. In the crab model, consolidation has been characterized in many aspects such as macromolecule synthesis, kinases, neurotransmitter receptors and transcription factors activation [59, 60, 93, 94, 64, 88, 95]. Furthermore, it was found that after re-exposure of the animals to the training context a new period of lability is opened in which a protein synthesis inhibitor and a NMDA receptor antagonist impairs LTM [88, 96]. Considering that NF-κB is necessary for consolidation in the crab, we tested whether this TF is critically required after memory reactivation, in keeping with the view that basic molecular mechanisms of consolidation are necessary to re-store the reactivated memory. NF-κB was specifically re-activated in animals re-exposed 24 h after training to the same training context but not to a different context. Furthermore, NF-κB was not activated in animals re-exposed to the context after a weak training protocol insufficient to induce long-term memory. Sulfasalazine impaired memory when administered 20 min before re-exposure to the training context but was not effective when a different context was used [97]. In relation to these findings, water shortage and angiotensin II administration, two treatments that activate NF-κB in this model, enhance long-term memory when presented simultaneously to memory reactivation [98]. These findings support that NF-κB is activated specifically by retrieval and that this activation, as well as in the initial consolidation phase, is required for memory stabilization after reactivation.

In rodents, inhibitory avoidance and fear conditioning are contextual memories in which the hippocampal formation plays a critical role. In both memory tasks, NF-κB is involved in reconsolidation. We found that the inhibition of NF-κB in hippocampus by sulfasalazine or decoy after memory reactivation impaired retention of inhibitory avoidance in mice. In contrast, a one base mutated decoy had no effect. Furthermore, we found NF-κB activation in the hippocampus, with a peak 15 min after memory retrieval [73]. This activation was faster than the one previously found in the hippocampus during consolidation of the same task, with a peak at 45 min [72], suggesting a temporal signature for both processes. The same activation kinetics was found in hippocampus in mice fear conditioning (de la Fuente and Romano, submitted). In other work, [14] found that retrieval of contextual conditioned fear memory activates the NF-κB pathway and that the mechanism is involved in chromatin regulation by histone H3 phosphorylation and acetylation at specific gene promoters in the hippocampus.

NF-κB IN EXTINCTION MEMORY

Inhibition of NF-κB During the Formation of Extinction Memory

In consolidated associative memories, retrieval induced by CS presentation without reinforcement may induce two apparently competing processes, reconsolidation and extinction. In contextual memories, a brief re-exposure to the training context induces memory reactivation while prolonged re-exposure induces extinction. A body of evidence supports that the original memory is not canceled by extinction, but its expression is temporarily inhibited [99]. The fact that extinction requires protein synthesis during a defined period of time led to the assumption that it entails a new memory that must consolidates as the original one. However, the temporary nature of extinction point to mechanistic differences with the original memory consolidation. In fact, recent studies support distinct molecular requirements, as the participation of protein phosphatases and endocanabinoids [100, 101].

As mentioned in the previous sections of this chapter, in the crab context-signal memory the activation of NF-κB plays a critical role in consolidation and reconsolidation. In this task, the re-exposure to the context without reinforcement can lead to reconsolidation of the original memory or to the formation of an extinction memory, depending on the duration of a single event of context re-exposure [96]. We studied the participation of NF-κB in memory extinction, and we found that the administration of the NF-κB inhibitor sulfasalazine prior to the extinction session impeded/delayed spontaneous recovery. Moreover, reinstatement experiments showed that the original memory was not affected and that NF-κB inhibition by sulfasalazine impaired or delayed spontaneous recovery, thus strengthening the ongoing memory extinction process. Interestingly, in animals with fully consolidated memory, a brief re-exposure to the training context induced NF-κB activation in the brain, while prolonged re-exposure induced NF-κB inhibition in correlation with memory extinction (Fig. **3**) [102]. We obtained similar results in ongoing experiments about the role of NF-κB in hippocampus in mice contextual fear conditioning (de la Fuente and Romano, submitted). These data represent a novel insight into the molecular mechanisms involved in the switch between memory reconsolidation and extinction.

On the basis of these findings, we propose a working model for NF-κB role in memory after retrieval (Fig. **4**). The initial process of transcriptional activation induced by retrieval would be mediated by protein kinases. In particular, the activation of IKK and PKA protein kinases induces NF-κB translocation to the nucleus and its activation. The prolonged presence of the CS would induce activation of other mediators such as protein phosphatases (i.e. calcineurin) that can increase the level of NF-κB inhibitor IκB and induce its nuclear exportation [103]. Under this interpretation, the administration of NF-κB inhibitors during memory reactivation reinforces the effect of prolonged exposure to the CS, provoking extinction strengthening.

Our interpretation is in line with the point of view that extinction formation recruited different mechanisms than the original memory consolidation and that weakening of the original consolidated circuit is part of the neural correlate of memory extinction. However, we cannot exclude the requirement of reinforcement mechanisms in other circuits, nondependent of NF-κB pathway, mediating the same extinction process. In summary, the evidence revised here supports that extinction does not require the activity of NF-κB, a key TF involved in consolidation and reconsolidation, but actually requires its inhibition.

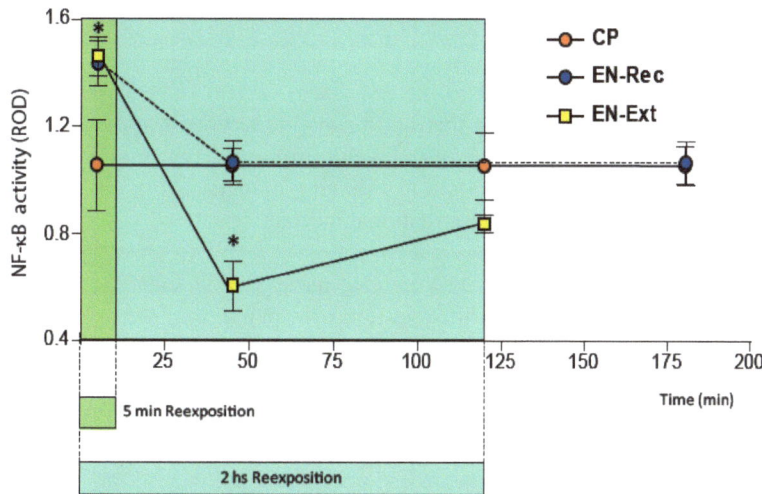

Fig. (3). Time course of NF-κB activity during memory reconsolidation and extinction. CT, control group; TR-Rec, trained group re-exposed for 5 minutes to the training context and sacrificed immediately, 45 min or 180 min after re-exposure; TR-Ext, trained group re-exposed 5, 45 or 120 minutes to the training context. *, $p < 0.05$ in T test. Figure redrawn from [102].

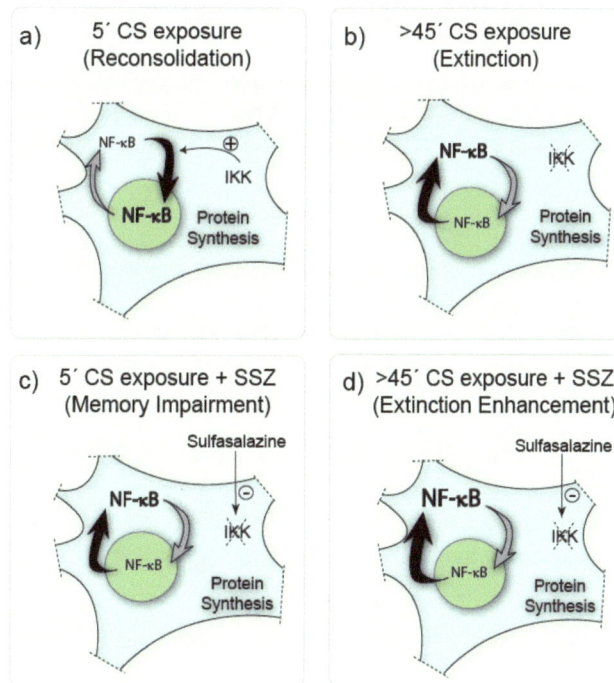

Fig. (4). A model of NF-κB role in memory reconsolidation and extinction. **a)** Five min context re-exposure induces NF-κB-dependent memory reconsolidation. **b)** Prolonged re-exposure induces NF-κB inhibition and memory extinction. **c)** NF-κB inhibition by sulfasalazine plus 5 min re-exposure impairs memory. **d)** NF-κB inhibition by sulfasalazine plus prolonged re-exposure induces extinction facilitation. Figure redrawn from [102].

CONCLUSIONS

A decade after the first description of the NF-κB participation in memory, a growing body of evidence supports an important role of this TF in the mechanisms of initial memory storage and in the processing after retrieval and re-activation. Such a role is evolutionarily conserved in vertebrates and invertebrates. In the formation of the extinction memory, very recent results resumed here support that the inhibition of NF-κB is required. This finding is supported by the fact that NF-κB inhibitory agents administered during extinction memory formation enhances long-term extinction. We propose that inhibition of transcription factors required for the original memory consolidation is necessary in the transition between memory reconsolidation and extinction.

Notwithstanding the development of the research in the last years, several critical issues remain to be elucidated. Perhaps the most important one is the presence of this TF in synaptic terminals and dendrites, which has led to the hypothesis of a synapse-to-nucleus signaling mechanism. We are in need of direct evidence for NF-κB retrograde-transport to the nucleus during memory consolidation in behaving animals. At this point other roles of NF-κB in synaptic terminals are also conceivable, related to translational mechanisms.

Identifying the target genes for NF-κB also remains an ongoing quest. In relation to this issue, the two waves of NF-κB activity observed during consolidation lead to the question of which genes have a fast and transient expression, which ones are expressed several hours later, and what is the role of each group of genes. Among the genes whose expression NF-κB contributes to regulate, four are central in neuroplasticity, the neurotrophic factor BDNF, the adhesion molecule NCAM, the IEG ZIF268 and the glutamate AMPA receptor (AMPAR). The AMPAR trafficking in synaptic membranes is considered a key mechanism in the activity dependent bi-directional plasticity of excitatory synapses [104].

The kinetic of NF-κB activation in mouse hippocampus in consolidation is different than in reconsolidation, both in inhibitory avoidance and fear conditioning. In consolidation the activity peaks appear at 45 min while in reconsolidation the time course is faster, reaching a peak at 15 min [72, 73]. One interpretation for the faster activation in reconsolidation suggests that molecular pathways leading to NF-κB nuclear translocation and DNA binding activity are facilitated by training. Another interpretation supports that, due to the initial consolidation process, hippocampal circuits are activated faster in retrieval and then, plasticity mechanisms such as NF-κB activation are recruited earlier. Beyond the interpretation for the faster curve in reconsolidation, these temporal differences imply a molecular signature for consolidation and reconsolidation that can be used as a tool to define whether new information may be incorporated in reconsolidation or if always required a new consolidation process [105].

Another point of interest is the ability of NF-κB to bind to CBP, which depends on p65 phosphorylation by PKA [70], the cAMP dependent protein kinase which has been implicated in several models of memory. The double activation of NF-κB by IKK and PKA pathways suggested the possibility of a Hebbian coincidence detector mechanism [106] (Fig. **3**). These particular features of the NF-κB pathway would enable the possibility that coincidence detection take place in synaptic compartment, allowing for a precise temporal and spatial determination of signals involved in neuronal plasticity and memory.

ACKNOWLEDGEMENTS

The work of my group was supported by ANPCYT PICT 26095, PICT 2049, and UBACYT X193 grants, Argentina. We especially thank Dr. Liliana Orelli for language correction and Martín Carbó Tanos for the illustrations.

REFERENCES

[1]   McGaugh JL. Time-dependent processes in memory storage. Science 1966; 153(742): 1351-8.
[2]   Davis HP, Squire LR. Protein synthesis and memory: a review. Psychol Bull 1984; 96(3): 518-59.

[3] Goelet P, Castellucci VF, Schacher S, Kandel ER. The long and the short of long-term memory--a molecular framework. Nature 1986; 322(6078): 419-22.

[4] Bliss CR, Sharp GW. A critical period in the development of the insulin secretory response to glucose in fetal rat pancreas. Life Sci 1994; 55(6): 423-7.

[5] Sutton MA, Schuman EM. Dendritic protein synthesis, synaptic plasticity, and memory. Cell 2006; 127(1): 49-58.

[6] Routtenberg A. The substrate for long-lasting memory: if not protein synthesis, then what? Neurobiol Learn Mem 2008; 89(3): 225-33.

[7] Routtenberg A, Rekart JL. Post-translational protein modification as the substrate for long-lasting memory. Trends Neurosci 2005; 28(1): 12-9.

[8] Matthies H. The biochemical basis of learning and memory. Life Sci 1974; 15(12): 2017-31.

[9] Matthies H. In search of cellular mechanisms of memory. Prog Neurobiol 1989; 32(4): 277-349.

[10] Rose SP. The biochemistry of memory. Essays Biochem 1991; 26: 1-12.

[11] Alberini CM. Genes to remember. J Exp Biol 1999; 202(Pt 21): 2887-91.

[12] Alberini CM. Transcription factors in long-term memory and synaptic plasticity. Physiol Rev 2009; 89(1): 121-45.

[13] Alberini CM, Ghirardi M, Metz R, Kandel ER. C/EBP is an immediate-early gene required for the consolidation of long-term facilitation in Aplysia. Cell 1994; 76(6): 1099-114.

[14] Lubin FD, Sweatt JD. The IkappaB kinase regulates chromatin structure during reconsolidation of conditioned fear memories. Neuron 2007; 55(6): 942-57.

[15] Kaang BK, Kandel ER, Grant SG. Activation of cAMP-responsive genes by stimuli that produce long-term facilitation in Aplysia sensory neurons. Neuron 1993; 10(3): 427-35.

[16] Yin JC, Tully T. CREB and the formation of long-term memory. Curr Opin Neurobiol 1996; 6(2): 264-8.

[17] Kandel ER. The molecular biology of memory storage: a dialog between genes and synapses. Biosci Rep 2004; 24(4-5): 475-522.

[18] Marini AM, Jiang X, Wu X, *et al.* Role of brain-derived neurotrophic factor and NF-kappaB in neuronal plasticity and survival: From genes to phenotype. Restor Neurol Neurosci 2004; 22(2): 121-30.

[19] Mellstrom B, Torres B, Link WA, Naranjo JR. The BDNF gene: exemplifying complexity in Ca2+ -dependent gene expression. Crit Rev Neurobiol 2004; 16(1-2): 43-9.

[20] Cavallaro S, Schreurs BG, Zhao W, D'Agata V, Alkon DL. Gene expression profiles during long-term memory consolidation. Eur J Neurosci 2001; 13(9): 1809-15.

[21] Cavallaro S, D'Agata V, Manickam P, Dufour F, Alkon DL. Memory-specific temporal profiles of gene expression in the hippocampus. Proc Natl Acad Sci USA 2002; 99(25): 16279-84.

[22] Levenson JM, Choi S, Lee SY, *et al.* A bioinformatics analysis of memory consolidation reveals involvement of the transcription factor c-rel. J Neurosci 2004; 24(16): 3933-43.

[23] Tian F, Hu X, Wu X, *et al.* Dynamic chromatin remodeling events in hippocampal neurons are associated with NMDA receptor-mediated activation of Bdnf gene promoter 1. J Neurochem 2009; 109: 1375-88.

[24] Clarke DL, Sutcliffe A, Deacon K, Bradbury D, Corbett L, Knox AJ. PKCbetaII augments NF-kappaB-dependent transcription at the CCL11 promoter via p300/CBP-associated factor recruitment and histone H4 acetylation. J Immunol 2008; 181(5): 3503-14.

[25] Saccani S, Pantano S, Natoli G. Two waves of nuclear factor kappaB recruitment to target promoters. J Exp Med 2001; 193(12): 1351-9.

[26] Levenson JM, Sweatt JD. Epigenetic mechanisms in memory formation. Nat Rev Neurosci 2005; 6(2): 108-18.

[27] Barrett RM, Wood MA. Beyond transcription factors: the role of chromatin modifying enzymes in regulating transcription required for memory. Learn Mem 2008; 15(7): 460-7.

[28] Ghosh S, Karin M. Missing pieces in the NF-kappaB puzzle. Cell 2002; 109 Suppl: S81-96.

[29] Sen R, Baltimore D. Inducibility of kappa immunoglobulin enhancer-binding protein Nf-kappa B by a posttranslational mechanism. Cell 1986; 47(6): 921-8.

[30] Meffert MK, Chang JM, Wiltgen BJ, Fanselow MS, Baltimore D. NF-kappa B functions in synaptic signaling and behavior. Nat Neurosci 2003; 6(10): 1072-8.

[31] O'Neill LA, Kaltschmidt C. NF-kappa B: a crucial transcription factor for glial and neuronal cell function. Trends Neurosci 1997; 20(6): 252-8.

[32] Frenkel L, Freudenthal R, Romano A, Nahmod VE, Maldonado H, Delorenzi A. Angiotensin II and the transcription factor Rel/NF-kappaB link environmental water shortage with memory improvement. Neuroscience 2002; 115(4): 1079-87.

[33] Krushel LA, Cunningham BA, Edelman GM, Crossin KL. NF-kappaB activity is induced by neural cell adhesion molecule binding to neurons and astrocytes. J Biol Chem 1999; 274(4): 2432-9.

[34] Choi J, Krushel LA, Crossin KL. NF-kappaB activation by N-CAM and cytokines in astrocytes is regulated by multiple protein kinases and redox modulation. Glia 2001; 33(1): 45-56.

[35] O'Mahony A, Raber J, Montano M, *et al.* NF-kappaB/Rel regulates inhibitory and excitatory neuronal function and synaptic plasticity. Mol Cell Biol 2006; 26(19): 7283-98.

[36] Lubin FD, Roth TL, Sweatt JD. Epigenetic regulation of BDNF gene transcription in the consolidation of fear memory. J Neurosci 2008; 28(42): 10576-86.

[37] Kaltschmidt C, Kaltschmidt B, Baeuerle PA. Brain synapses contain inducible forms of the transcription factor NF-kappa B. Mech Dev 1993; 43(2-3): 135-47.

[38] Guerrini L, Blasi F, Denis-Donini S. Synaptic activation of NF-kappa B by glutamate in cerebellar granule neurons in vitro. Proc Natl Acad Sci USA 1995; 92(20): 9077-81.

[39] Meberg PJ, Kinney WR, Valcourt EG, Routtenberg A. Gene expression of the transcription factor NF-kappa B in hippocampus: regulation by synaptic activity. Brain Res Mol Brain Res 1996; 38(2): 179-90.

[40] Hanz S, Perlson E, Willis D, *et al.* Axoplasmic importins enable retrograde injury signaling in lesioned nerve. Neuron 2003; 40(6): 1095-104.

[41] Thompson KR, Otis KO, Chen DY, Zhao Y, O'Dell TJ, Martin KC. Synapse to nucleus signaling during long-term synaptic plasticity; a role for the classical active nuclear import pathway. Neuron 2004; 44(6): 997-1009.

[42] Mikenberg I, Widera D, Kaus A, Kaltschmidt B, Kaltschmidt C. Transcription factor NF-kappaB is transported to the nucleus via cytoplasmic dynein/dynactin motor complex in hippocampal neurons. PLoS One 2007; 2(7): e589.

[43] Wellmann H, Kaltschmidt B, Kaltschmidt C. Retrograde transport of transcription factor NF-kappa B in living neurons. J Biol Chem 2001; 276(15): 11821-9.

[44] Bliss TV, Collingridge GL. A synaptic model of memory: long-term potentiation in the hippocampus. Nature 1993; 361(6407): 31-9.

[45]   Manahan-Vaughan D, Kulla A, Frey JU. Requirement of translation but not transcription for the maintenance of long-term depression in the CA1 region of freely moving rats. J Neurosci 2000; 20(22): 8572-6.

[46]   Lindecke A, Korte M, Zagrebelsky M, *et al.* Long-term depression activates transcription of immediate early transcription factor genes: involvement of serum response factor/Elk-1. Eur J Neurosci 2006; 24(2): 555-63.

[47]   Albensi BC, Mattson MP. Evidence for the involvement of TNF and NF-kappaB in hippocampal synaptic plasticity. Synapse 2000; 35(2): 151-9.

[48]   Yeh SH, Lin CH, Lee CF, Gean PW. A requirement of nuclear factor-kappaB activation in fear-potentiated startle. J Biol Chem 2002; 277(48): 46720-9.

[49]   Freudenthal R, Romano A, Routtenberg A. Transcription factor NF-kappaB activation after in vivo perforant path LTP in mouse hippocampus. Hippocampus 2004; 14(6): 677-83.

[50]   Igaz LM, Vianna MR, Medina JH, Izquierdo I. Two time periods of hippocampal mRNA synthesis are required for memory consolidation of fear-motivated learning. J Neurosci 2002; 22(15): 6781-9.

[51]   Dash PK, Hochner B, Kandel ER. Injection of the cAMP-responsive element into the nucleus of Aplysia sensory neurons blocks long-term facilitation. Nature 1990; 345(6277): 718-21.

[52]   Taubenfeld SM, Milekic MH, Monti B, Alberini CM. The consolidation of new but not reactivated memory requires hippocampal C/EBPbeta. Nat Neurosci 2001; 4(8): 813-8.

[53]   Lee JL, Di Ciano P, Thomas KL, Everitt BJ. Disrupting reconsolidation of drug memories reduces cocaine-seeking behavior. Neuron 2005; 47(6): 795-801.

[54]   Josselyn SA, Kida S, Silva AJ. Inducible repression of CREB function disrupts amygdala-dependent memory. Neurobiol Learn Mem 2004; 82(2): 159-63.

[55]   Merlo E, Freudenthal R, Romano A. The IkappaB kinase inhibitor sulfasalazine impairs long-term memory in the crab Chasmagnathus. Neuroscience 2002; 112(1): 161-72.

[56]   Lozada M, Romano A, Maldonado H. Long-term habituation to a danger stimulus in the crab Chasmagnathus granulatus. Physiol Behav 1990; 47(1): 35-41.

[57]   Pereyra P, Gonzalez Portino E, Maldonado H. Long-lasting and context-specific freezing preference is acquired after spaced repeated presentations of a danger stimulus in the crab Chasmagnathus. Neurobiol Learn Mem 2000; 74(2): 119-34.

[58]   Tomsic D, Romano A, Maldonado H. Behavioral and mechanistic bases of long-term habituation in the crab Chasmagnathus. Adv Exp Med Biol 1998; 446: 17-35.

[59]   Pedreira ME, Dimant B, Tomsic D, Quesada-Allue LA, Maldonado H. Cycloheximide inhibits context memory and long-term habituation in the crab Chasmagnathus. Pharmacol Biochem Behav 1995; 52(2): 385-95.

[60]   Pedreira ME, Dimant B, Maldonado H. Inhibitors of protein and RNA synthesis block context memory and long-term habituation in the crab Chasmagnathus. Pharmacol Biochem Behav 1996; 54(3): 611-7.

[61]   Pedreira ME, Romano A, Tomsic. D., Lozada M, Maldonado H. Massed and spaced training build up different components of long-term habituation in the crab Chasmagnathus. . Animal Learning and Behavior 1998; 26(1): 34-8.

[62]   Hermitte G, Pedreira ME, Tomsic D, Maldonado H. Context shift and protein synthesis inhibition disrupt long-term habituation after spaced, but not massed, training in the crab Chasmagnathus. Neurobiol Learn Mem 1999; 71(1): 34-49.

[63]   Freudenthal R, Locatelli F, Hermitte G, *et al.* Kappa-B like DNA-binding activity is enhanced after spaced training that induces long-term memory in the crab Chasmagnathus. Neurosci Lett 1998; 242(3): 143-6.

[64]   Freudenthal R, Romano A. Participation of Rel/NF-kappaB transcription factors in long-term memory in the crab Chasmagnathus. Brain Res 2000; 855(2): 274-81.

[65]   Hohmann HP, Remy R, Aigner L, Brockhaus M, van Loon AP. Protein kinases negatively affect nuclear factor-kappa B activation by tumor necrosis factor-alpha at two different stages in promyelocytic HL60 cells. J Biol Chem 1992; 267(3): 2065-72.

[66]   Hoult JR. Pharmacological and biochemical actions of sulphasalazine. Drugs 1986; 32 Suppl 1: 18-26.

[67]   Wahl C, Liptay S, Adler G, Schmid RM. Sulfasalazine: a potent and specific inhibitor of nuclear factor kappa B. J Clin Invest 1998; 101(5): 1163-74.

[68]   Weber CK, Liptay S, Wirth T, Adler G, Schmid RM. Suppression of NF-kappaB activity by sulfasalazine is mediated by direct inhibition of IkappaB kinases alpha and beta. Gastroenterology 2000; 119(5): 1209-18.

[69]   Yeh SH, Lin CH, Gean PW. Acetylation of nuclear factor-kappaB in rat amygdala improves long-term but not short-term retention of fear memory. Mol Pharmacol 2004; 65(5): 1286-92.

[70]   Zhong H, Voll RE, Ghosh S. Phosphorylation of NF-kappa B p65 by PKA stimulates transcriptional activity by promoting a novel bivalent interaction with the coactivator CBP/p300. Mol Cell 1998; 1(5): 661-71.

[71]   Boccia MM, Acosta GB, Blake MG, Baratti CM. Memory consolidation and reconsolidation of an inhibitory avoidance response in mice: effects of i.c.v. injections of hemicholinium-3. Neuroscience 2004; 124(4): 735-41.

[72]   Freudenthal R, Boccia MM, Acosta GB, *et al.* NF-kappaB transcription factor is required for inhibitory avoidance long-term memory in mice. Eur J Neurosci 2005; 21(10): 2845-52.

[73]   Boccia M, Freudenthal R, Blake M, *et al.* Activation of hippocampal nuclear factor-kappa B by retrieval is required for memory reconsolidation. J Neurosci 2007; 27(49): 13436-45.

[74]   Kassed CA, Willing AE, Garbuzova-Davis S, Sanberg PR, Pennypacker KR. Lack of NF-kappaB p50 exacerbates degeneration of hippocampal neurons after chemical exposure and impairs learning. Exp Neurol 2002; 176(2): 277-88.

[75]   Kassed CA, Herkenham M. NF-kappaB p50-deficient mice show reduced anxiety-like behaviors in tests of exploratory drive and anxiety. Behav Brain Res 2004; 154(2): 577-84.

[76]   Beg AA, Baltimore D. An essential role for NF-kappaB in preventing TNF-alpha-induced cell death. Science 1996; 274(5288): 782-4.

[77]   Alcamo E, Mizgerd JP, Horwitz BH, *et al.* Targeted mutation of TNF receptor I rescues the RelA-deficient mouse and reveals a critical role for NF-kappa B in leukocyte recruitment. J Immunol 2001; 167(3): 1592-600.

[78]   Kaltschmidt B, Ndiaye D, Korte M, *et al.* NF-kappaB regulates spatial memory formation and synaptic plasticity through protein kinase A/CREB signaling. Mol Cell Biol 2006; 26(8): 2936-46.

[79]   Zhou LJ, Zhong Y, Ren WJ, Li YY, Zhang T, Liu XG. BDNF induces late-phase LTP of C-fiber evoked field potentials in rat spinal dorsal horn. Exp Neurol 2008; 212(2): 507-14.

[80]   Delorenzi A, Pedreira ME, Romano A, *et al.* Angiotensin II enhances long-term memory in the crab Chasmagnathus. Brain Res Bull 1996; 41(4): 211-20.

[81]   Delorenzi A, Dimant B, Frenkel L, Nahmod VE, Nassel DR, Maldonado H. High environmental salinity induces memory enhancement and increases levels of brain angiotensin-like peptides in the crab Chasmagnathus granulatus. J Exp Biol 2000; 203(Pt 22): 3369-79.

[82]   Bekinschtein P, Cammarota M, Igaz LM, Bevilaqua LR, Izquierdo I, Medina JH. Persistence of long-term memory storage requires a late protein synthesis- and BDNF- dependent phase in the hippocampus. Neuron 2007; 53(2): 261-77.

[83]   Rossato JI, Bevilaqua LR, Izquierdo I, Medina JH, Cammarota M. Dopamine controls persistence of long-term memory storage. Science 2009; 325(5943): 1017-20.

[84]   Misanin JR, Miller RR, Lewis DJ. Retrograde amnesia produced by electroconvulsive shock after reactivation of a consolidated memory trace. Science 1968; 160(827): 554-5.

[85]   Mactutus CF, Riccio DC, Ferek JM. Retrograde amnesia for old (reactivated) memory: some anomalous characteristics. Science 1979; 204(4399): 1319-20.

[86]   Sara SJ. Retrieval and reconsolidation: toward a neurobiology of remembering. Learn Mem 2000; 7(2): 73-84.

[87]   Nader K, Schafe GE, Le Doux JE. Fear memories require protein synthesis in the amygdala for reconsolidation after retrieval. Nature 2000; 406(6797): 722-6.

[88]   Pedreira ME, Perez-Cuesta LM, Maldonado H. Reactivation and reconsolidation of long-term memory in the crab Chasmagnathus: protein synthesis requirement and mediation by NMDA-type glutamatergic receptors. J Neurosci 2002; 22(18): 8305-11.

[89]   Debiec J, LeDoux JE, Nader K. Cellular and systems reconsolidation in the hippocampus. Neuron 2002; 36(3): 527-38.

[90]   Myers KM, Davis M. Behavioral and neural analysis of extinction. Neuron 2002; 36(4): 567-84.

[91]   Lee JL, Everitt BJ, Thomas KL. Independent cellular processes for hippocampal memory consolidation and reconsolidation. Science 2004; 304(5672): 839-43.

[92]   Salinska E, Bourne RC, Rose SP. Reminder effects: the molecular cascade following a reminder in young chicks does not recapitulate that following training on a passive avoidance task. Eur J Neurosci 2004; 19(11): 3042-7.

[93]   Locatelli F, Maldonado H, Romano A. Two critical periods for cAMP-dependent protein kinase activity during long-term memory consolidation in the crab Chasmagnathus. Neurobiol Learn Mem 2002; 77(2): 234-49.

[94]   Locatelli F, Romano A. Differential activity profile of cAMP-dependent protein kinase isoforms during long-term memory consolidation in the crab Chasmagnathus. Neurobiol Learn Mem 2005; 83(3): 232-42.

[95]   Feld M, Dimant B, Delorenzi A, Coso O, Romano A. Phosphorylation of extra-nuclear ERK/MAPK is required for long-term memory consolidation in the crab Chasmagnathus. Behav Brain Res 2005; 158(2): 251-61.

[96]   Pedreira ME, Maldonado H. Protein synthesis subserves reconsolidation or extinction depending on reminder duration. Neuron 2003; 38(6): 863-9.

[97]   Merlo E, Freudenthal R, Maldonado H, Romano A. Activation of the transcription factor NF-kappaB by retrieval is required for long-term memory reconsolidation. Learn Mem 2005; 12(1): 23-9.

[98]   Frenkel L, Maldonado H, Delorenzi A. Memory strengthening by a real-life episode during reconsolidation: an outcome of water deprivation via brain angiotensin II. Eur J Neurosci 2005; 22(7): 1757-66.

[99]   Rescorla RA. Spontaneous recovery. Learn Mem 2004; 11(5): 501-9.

[100]   Genoux D, Haditsch U, Knobloch M, Michalon A, Storm D, Mansuy IM. Protein phosphatase 1 is a molecular constraint on learning and memory. Nature 2002; 418(6901): 970-5.

[101]   Suzuki A, Josselyn SA, Frankland PW, Masushige S, Silva AJ, Kida S. Memory reconsolidation and extinction have distinct temporal and biochemical signatures. J Neurosci 2004; 24(20): 4787-95.

[102]   Merlo E, Romano A. Memory extinction entails the inhibition of the transcription factor NF-kappaB. PLoS One 2008; 3(11): e3687.

[103]   Arenzana-Seisdedos F, Turpin P, Rodriguez M, *et al.* Nuclear localization of I kappa B alpha promotes active transport of NF-kappa B from the nucleus to the cytoplasm. J Cell Sci 1997; 110 (Pt 3): 369-78.

[104]   Citri A, Malenka RC. Synaptic plasticity: multiple forms, functions, and mechanisms. Neuropsychopharmacology 2008; 33(1): 18-41.

[105]   Tronel S, Milekic MH, Alberini CM. Linking new information to a reactivated memory requires consolidation and not reconsolidation mechanisms. PLoS Biol 2005; 3(9): e293.

[106]   Barger SW, Moerman AM, Mao X. Molecular mechanisms of cytokine-induced neuroprotection: NFkappaB and neuroplasticity. Curr Pharm Des 2005; 11(8): 985-98.

CHAPTER 7

NF-κB in Neurons—Mechanisms and Myths

Steven W. Barger[1,2,3]* and Xianrong R. Mao[4]

[1]Department of Geriatrics, [2]Department of Neurobiology and Developmental Sciences, University of Arkansas for Medical Sciences, Little Rock AR 72205, USA; [3]Geriatric Research Education and Clinical Center, Central Arkansas Veterans Healthcare System, Little Rock AR 72205; [4]Department of Anesthesiology, Washington University School of Medicine, St. Louis MO USA 63110

Abstract: The transcriptional activation of specific genes by transcription factor proteins is an important factor in the determination of cell-type specific patterns of gene expression. In one form or another, transcription factors comprising the Rel-family of proteins (responsible for the activity referred to as "NF-κB") are present in every cell type examined. However, studies of NF-κB often rely solely on a single endpoint (e.g., nuclear translocation) as an index of activation. Careful examination of CNS neuronal populations indicates that the initial components of NF-κB activation, up to and including nuclear translocation, are often dissociated from transcriptional activation. Indeed, there are few, if any, circumstances in which classical transcriptional activation by NF-κB has been documented in CNS neurons. In addition to the mechanistic intrigue this disjunction inspires, it is possible that this phenomenon contributes to important cell-type specificity distinguishing neurons from other cell types. It also suggests several implications for pharmacotherapeutic manipulation of NF-κB in the CNS.

Keywords: Artifact, cell culture, Glia, neurons, nuclear factor kappa B, rel family, reporter gene, specificity protein transcription factor (Sp1), specificity protein 4 transcription factor (Sp4), transcription.

INTRODUCTION TO NF-KB AND ITS ACTIVATION

NF-κB transcription factors are critically involved in immune function and several developmental events (reviewed in [1, 2]). The genes most robustly and consistently activated by NF-κB are those responsible for enzymes (e.g., nitric oxide synthase-2) or cytokines (e.g., tumor necrosis factor-α and interleukin-6) contributing to inflammation. Apropos the urgency of mounting defenses against fast-growing pathogens, NF-κB can be activated rapidly by virtue of the fact that many cells contain substantial amounts of this transcription factor, held "at the ready" in their cytosol. Translocation to the nucleus is all that is required for activation in many cases. Because of this potential for rapid responses and other nuances of activation discussed below, neuroscientists have speculated that NF-κB might contribute to transcriptional events linking neurochemistry to neuroplasticity. While evidence of a requirement for NF-κB in some neuroplasticity paradigms is intriguing, exclusion of artifacts from the canon of literature leaves no appreciable evidence that this transcription factor acts within excitatory synaptic potentiation as first suggested by the earliest studies. Indeed, the record of investigation into this hypothetical role for NF-κB serves as a cautionary tale regarding the challenges that neurophysiology can pose for researchers.

Structurally, active NF-κB consists of a homo- or heterodimer of members of the Rel protein family. The diversity in this family and the potential for multiple dimeric combinations creates some variety in the DNA sequences to which these proteins bind. However, there are limits—not every member of the Rel family can dimerize with every other member. The major dimers in vertebrates include RelA/p50, RelB/p52, c-Rel/RelA as well as homodimers of RelA, p50, and c-Rel. The consensus sequence that was established for binding the most prominent form of NF-κB

*Address correspondence to Steven W. Barger:** Department of Geriatrics, Department of Neurobiology and Developmental Sciences, University of Arkansas for Medical Sciences, Little Rock AR, 72205 USA and Geriatric Research Education and Clinical Center, Central Arkansas Veterans Healthcare, System, Little Rock AR 72205 USA; E-mail: BargerStevenW@uams.edu

(RelA/p50 heterodimer) is GGGRNNYYCC. Within this consensus, "Rel" proteins (RelA, RelB, c-Rel) bind the 3' "half site" and p50 (or p52) binds the 5' [3].

The various NF-κB moieties are present in an inactive state in the cytosol (and mitochondria [4]) of most cell types, held latent by the binding of one of various inhibitory proteins collectively termed IκBs. Signal transduction events leading to activation of NF-κB result in degradation of the IκB, typically by the ubiquitin-proteasome system. This frees the active NF-κB dimer to translocate to the nucleus, bind specific DNA sequences (the κB enhancer), and recruit the transcriptosome. Owing to its ability to inhibit NF-κB's DNA-binding activity, mask its nuclear localization signal, and present an efficient nuclear export signal, the α-isoform of IκB also participates in rapid reversals of NF-κB activation, and this phenomenon contributes to a responsive feedback control aided by the fact that the promoter of IκBα itself contains a κB enhancer [5].

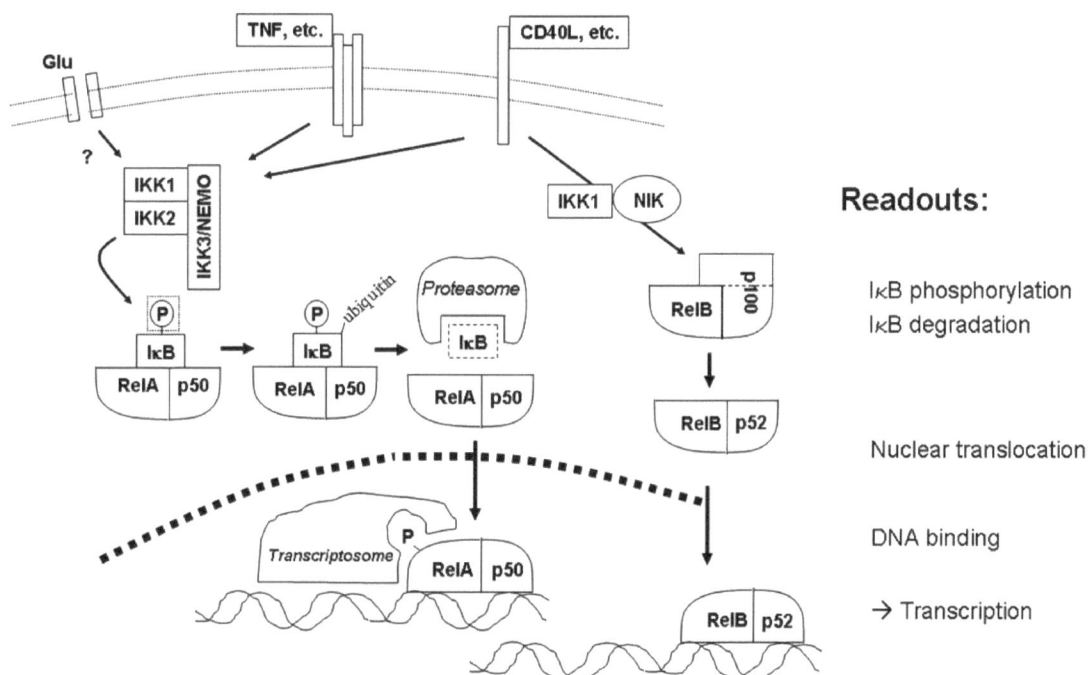

Fig. (1). NF-κB activation—correspondence to methodological endpoints. The activation of NF-κB is typically triggered by ligation of plasmalemmal receptors, some of which (e.g., TNFα receptors) are specifically coupled to activation of a complex containing IKK1, IKK2, and IKK3/NEMO in the canonical pathway. This leads to phosphorylation of IκBs (typically, IκBα and IκBβ), targeting them for ubiquitination and degradation by the proteasome. This unmasks the nuclear localization signal of canonical NF-κB dimers (RelA/RelA, RelA/p50, RelA/c-Rel) such that they translocate to the nucleus. Other plasmalemmal receptors can specifically activate NIK and IKK1 in the noncanonical pathway (e.g., lymphotoxin-α receptors) or both pathways (e.g., CD40). This results in phosphorylation-triggered ubiquitination of p100, which serves as both an IκB and the precursor for p52. Proteolytic removal of the C-terminal half of p100 leaves p52 bound to RelB, and together, this heterodimer comprises noncanonical NF-κB, which also translocates to the nucleus. Questionable claims have been made that ionotropic glutamate receptors activate RelA-containing NF-κB, i.e., the canonical pathway. Once in the nucleus, NF-κB binds to promoter elements that usually satisfy the consensus sequence GGGRNNYYCC. In some cases, phosphorylation of the transactivating subunit (RelA, RelB, or c-Rel) seems to be required for recruitment of the transcriptosome and initiation of RNA polymerase II activity. NF-κB research projects variably assay these steps through the endpoint readouts listed to the right, utilizing the techniques of phospho-specific western blot analysis, immunocytochemistry, EMSA, ChIP assays, and reporter-gene assays. A single research project rarely applies all of these methods as we have in testing the glutamate-triggered activation of NF-κB in CNS neurons. Only the nuclear translocation of NF-κB has been documented in this system.

The transactivating moieties of NF-κB include heterodimers of RelA (p65) with c-Rel or, more prevalently, p50; homodimers of RelA or c-Rel; or heterodimers of RelB with p52. The latter is activated by a signal transduction scheme that differs from the others and is thus termed "noncanonical" or "alternative" NF-κB (reviewed in [6]). The others are usually referred to as "canonical" NF-κB and are activated by IKK2 operating in conjuction with IKK3, though only the IKK2 appears to contibute kinase activity. It phosphorylates critical serine residues on IκBα or IκBβ associated with the canonical dimers, marking these IκBs for ubiquitination and proteasomal degradation (Fig. **1**). Activation of noncanonical NF-κB (RelB/p52) occurs somewhat differently. The latent moeity is RelB complexed with the precursor of p52, p100. Activation is triggered by NF-κB-inducing kinase (NIK, aka MAP3K14), which activates phosphorylation of p100 by IKK1. Serine phosphorylation of the p100 subunit marks the protein for cleavage into p52. This noncanonical pathway is activated by specific receptors and thus is employed only by specific cytokine signals, lymphotoxin-α1β2 and CD40 ligand among them. Noncanonical activity tends to build more slowly and remain sustained several hours longer than does the activation of canonical NF-κB [7, 8].

There are other, unconventional routes to NF-κB activation which have occasionally been termed "noncanonical," leading to some consternation among those who reserve this term for the activation of RelB/p52. Among these other mechanisms is the tyrosine phosphorylation of IκBα [12]. IKK2 working through the canonical pathway phosphorylates serines (32 and 36) of IκBα; thus, mutagenesis of these residues to alanine creates an IκBα that cannot be degraded and thus acts as a transdominant repressor (sometimes referred to as a "super repressor"). But there are certain paradigms in which this transdominant repressor can be bypassed. One appears to be the notable ability of oxidative stress to induce NF-κB. Takada *et al.* ([13]) reported that hydrogen peroxide stimulates tyrosine phosphorylation of IκBα through Syk kinase. Tyrosine phosphorylation of IκBα does not appear to mark it for proteosomal degradation the way IKK2 phosphorylation of Serines-32/-36 does; rather, it simply causes dissociation from the RelA/p50 or c-Rel complexes, permitting these to translocate to the nucleus. Similar activations of NF-κB through tyrosine phosphorylation of IκBα have been reported for nerve growth factor and ciliary neurotrophic factor [14, 15].

Qin *et al.* [16] have concluded that excitatory amino acids activate NF-κB in the striatum *via* a mechanism that is independent of the proteasome but dependent upon caspase-3. This was based on injections of Ac-DEVD-CHO prior to quisqualate in the striatum. Though selective for caspase-3 at low concentrations, Ac-DEVD-CHO can inhibit other enzymes. Even if the drug rapidly diffused throughout the entire brain, the doses applied by Qin *et al.* may have achieved 8 micromolar, a concentration 400-fold above its K_i for caspase-1. Caspase-1 (aka, interleukin-1β-converting enzyme) is necessary for the production of IL-1β, a potent inducer of NF-κB. Thus, the effects of Ac-DEVD-CHO may be explained *via* NF-κB's tried-and-true connection to inflammatory cytokines rather than a novel caspase-3 attack on IκB.

In addition to these schema for activation of NF-κB-responsive genes, there is another pathway that involves p105. This p50 precursor appears to act as a tonic inhibitor of "tumor progression locus-2" (TPL-2, aka MAP3K8), a kinase capable of activating MEK and, by extension, ERK kinases. Regulated degradation of p105 results in liberation of TPL-2 and activation of MEK→ERK [17]. Interestingly, p105 sequestration not only inhibits the kinase activity of TPL-2, it also stabilizes it. Thus, p105-deficient cells show drastically attenuated ERK activation in response to NF-κB-activating stimuli that happen to degrade p105, such as Toll-like receptor (TLR) ligands (e.g., lipopolysaccharide, gram-positive lipoproteins, double-stranded RNA, unmethylated CpG-containing DNA, etc.) [17]. This has consequences for a popular experimental tool: many studies have attempted to draw conclusions about roles of NF-κB in various systems by studying p105-knockout mice. In addition to the unavoidable caveat (discussed further below) that this manipulation removes the bioactivity of p105, p50 homodimers, and IκBγ [18], it also drastically attentuates signal transduction involving TPL-2 and MEK.

PHOSPHORYLATION OF RELA

Excitatory amino acids could also impact NF-κB activity through the nitric oxide (NO)→cyclic GMP pathway. In several types of CNS neurons, notably cerebellar granule and basket cells, glutamate receptor agonism is associated with a calcium-dependent activation of the calmodulin-dependent enzyme neuronal NO synthase (nNOS). The NO produced activates soluble guanylate cyclase to produce cGMP which, in turn, activates cGMP-dependent protein kinase (PKG). In T-cells, PKG phosphorylates RelA, p50, and p52 [19]. RelA thus modified induce transcription from conventional κB elements. Moreover, phosphorylation of p50 and p52 stimulates their ability to participate in transactivation at an enhancer element that consists of a C/EBP site and a p50 half site (above). It should be emphasized, however, that these PKG-dependent events have not been documented in response to excitatory amino acids specifically nor, for that matter, in neurons generally.

The nuances of PKG notwithstanding, considerable evidence indicates that phosphorylation of RelA is either necessary for transactivation or can otherwise alter the outcome of NF-κB activation. Sites of RelA phosphorylation and their attendant kinases include Serine-276 by protein kinase A [20, 21], Serine-471 by PKC-ε [22-24], Serine-529 by casein kinase II [25-27], and Serine-536 by IKK-2 [28-31]. RelA also appears to be phosphorylated somewhere in its transactivation domain by Ca^{2+}/calmodulin-dependent protein kinase IV [32] and glycogen synthase kinase-3β [33]. In some systems RelA has been reported to be transcriptionally impotent unless phosphorylated at specific positions, but in other reports this requirement is less rigid; yet phosphorylation may effect subtler changes in the level of activity or the DNA target site RelA prefers. In superior cervical ganglion sympathetic neurons and nodose ganglion sensory neurons, NF-κB activation inhibits neurite outgrowth unless phosphorylation of RelA (at Serine-536) is blocked, in which case NF-κB activity is associated with enhanced neuritogenesis [34]. This may involve changes in RelA's preferred target sequence—and, thus, the gene promoter that it regulates—resulting from phosphorylation of Serine-536, as documented in T-cells [35].

ACETYLATION

In addition to phosphorylation, another post-translational modification of NF-κB subunits contributes to the control of function: RelA is acetylated on several lysine residues (specifically, Lysine-122, -123, -218, -221, and -310) in a regulated, reversible manner [36-40]. The responsible enzyme is p300/CBP, and its acetylation of RelA modifies NF-κB activity [37, 39, 41]. Acetylation of lysine-221 enhances its DNA-binding properties and, together with the acetylation of lysine-218, impairs the assembly of RelA with IκBα. Acetylation of lysine-310 of RelA has no apparent effect on DNA binding activity but modifies transcriptional transactivation activity [37] Mutation of Lysine-310 in RelA to arginine attenuates the transactivation of NF-κB and suppresses expression of cytokines [37, 38, 40].

METHODS APPLIED TO MONITOR NF-κB

The mode of activation of NF-κB lends itself to several experimental assays that have been applied to studies of this transcription factor. As outlined in Fig. **1**, various techniques make it possible to measure the phosphorylation of the IκBs, their degradation, translocation of the active dimers to the nucleus, DNA binding, and transcriptional activation. Western immunoblot analysis has been useful for monitoring IκB phosphorylation and degradation, as well as nuclear translocation after cell fractionation. The latter can also be surveyed by immunocytochemistry. A significant advance was claimed when antibodies were developed against the nuclear localization signal (NLS) of RelA [42]. Because this domain is masked by IκB binding, it was reasoned that changes in immunodetection by this type of antibody would be a reliable readout of activation. However, there are additional requirements for transcriptional activation by NF-κB; thus, unmasking of the NLS—or even nuclear translocation—is not tantamount to activation (below).

Binding to a specific DNA sequence is an essential aspect of a transcription factor's activity and thus serves as an important endpoint in studies of NF-κB. The majority of studies have relied upon electrophoretic mobility shift assay (EMSA) for this determination. Commonly referred to as "gel shift," EMSA uses a labeled nucleic acid probe—in this case, dsDNA containing an NF-κB target sequence. The probe is incubated with protein extracts, typically those obtained from the nuclei of cell cultures or tissue preparations. The incubation mixture is then resolved on a nondenaturing polyacrylamide gel. The unbound probe migrates fast, whereas probe bound by protein is retarded; hence, "shifted." Specificity is ensured by the inclusion of nonspecific DNA or other similar polymers, as well as control reactions that contain an excess of unlabeled probe or labeled mutated probe. EMSA may be augmented by "supershift," effected by adding into the probe/extract mix an antibody to an NF-κB subunit. Because EMSA gels are run under native conditions, the antibody stays associated with the transcription factor, thereby creating a larger quaternary structure and further retarding the migration of the protein/DNA complex. This is the only efficient way to confirm identity of the protein responsible for the shift in EMSA.

Chromatin immunoprecipitation (ChIP) assay is another techique used to assay DNA binding. In this assay, cells are fixed in formaldehyde to lock the tertiary/quaternary structure of transcription factors, some of which are bound to their specific DNA targets. The cellular DNA is then broken into ~200-basepair fragments, typically by sonication, and the DNA sequences complexed with a specific transcription factor are isolated by immunoprecipitating the protein. Heating the precipitates relaxes the proteins enough to liberate the DNA, which can then be identified and quantified by polymerase chain reaction. Because of the dependency on immunoprecipitation, ChIP assays are inherently more specific for the protein of interest than is EMSA. However, the relatively long stretch of DNA isolated for PCR amplification means that it is difficult to ascertain the precise region of DNA responsible for interactions with protein. In addition, it is possible for the results of ChIP to be complicated by protein-protein interactions; the protein recognized by the antibody may be associated with the DNA indirectly *via* binding to another DNA-binding protein rather than to the DNA itself.

Although there are intriguing alternatives, transcriptional activation is the endpoint most relevant to the study of NF-κB's role in cellular physiology. This can only be defensibly measured as the expression of an NF-κB-regulated gene, whether it is an endogenous gene or a transgenic reporter gene such as luciferase or β-galactosidase. The latter affords the opportunity to simplify the promoter to make it more specifically dependent on NF-κB (whereas an endogenous gene may have several different regulatory cis elements that respond to sundry transcription factors). It also as the advantage of easy mutation so that comparisons can be made for the purpose of verifying specificity. One disadvantage is that the many copies of a reporter plasmid that typically accumulate in a transfected cell can interrupt the stoichiometry and therefore the function of the transcription factor(s) of interest. One can even achieve such a high copy number that the transcription factors binding the plasmid are diluted out to the point of ineffectiveness.

Although the steps in NF-κB activation typically occur together, many instances have been reported of the initial stages being instigated without a resultant change in transcription. Nuclear translocation has been a particularly frustrating red herring; many paradigms have been reported in which NF-κB shifts to the nucleus without stimulating transcription [27, 43-50] Wang, 1998 #2650. This may indicate roles for NF-κB other than straightforward, classical transactivation; e.g., it may function in corepressor or transrepressor schema [51, 52]. At the very least, however, it points up the danger in drawing conclusions from assays of nuclear translocation alone.

EXPERIMENTAL MANIPULATION OF NF-κB

In order to test the role of NF-κB in a given system, the most rigorous approach is to examine outcomes after specifically inhibiting its activity or removing components of its subunits. Of course, this also has clinical relevance in circumstances where NF-κB is hyperactivated and contributing to inflammatory or neoplastic diseases. Indeed, evidence suggests that inhibition of NF-κB is an important auxiliary activity of several pharmacological agents that have another primary target. Thus, some of the anti-inflammatory effects of aspirin and other salicylates probably results from their inhibition of NF-κB [53]. This may also be true of certain agonists of peroxisome proliferator-activated receptor γ (PPARγ), sometimes in a PPARγ-independent manner [54-57]. Experimentally, such compounds have been used to argue for a role for NF-κB in various models. However, their utility in this regard is obviously compromised by the primary effects these agents have on the targets for which they were developed. Similar side effects plague experimental designs that make use of various antioxidants to inhibit NF-κB (an action due, most likely, to the general ability of cellular oxidation to inhibit IκBα as mentioned above), including pyrolidine dithiocarbamate (PDTC), butylated hydroxy anisole (BHA), and diethyldithiocarbamate (DETC). These agents are unfortunately nonspecific. PDTC, for instance, inhibits JNK; it also stimulates p38 and ERK MAP kinases [58]. DETC has severe neurotoxicity due in part to its inhibition of superoxide dismutase [59].

One of the early tools claimed to specifically inhibit NF-κB is a peptide based on the NLS of p50. The rationale is that this peptide—termed "SN50"—will compete with p50 or p50-containing complexes at the transporter responsible for transferring such a protein to the nucleus. Presumably, this peptide would also compete with any other protein that utilizes the same transporter (it is unlikely that p50, or indeed any single protein, would have its own unique transporter). And if it is specific for p50, this would make it less effective against forms of NF-κB other than the canonical RelA/p50 heterodimer. This conflict seems to be born out by a report demonstrating specificity of an analogous peptide based on the NLS of p52 ("SN52"), which had no discernible effect on the nuclear localization of p50's partner RelA [60]. The same report showed marginal crossover of SN50 in inhibition of RelB, again pointing up the intuitive idea that any such competitive inhibitor of a binding site would not be expected to be specific for a given ligand.

Another protocol involving intracellular delivery of a macromolecule stands to effect a more specific inhibition of NF-κB. Specifically, double-stranded DNA containing an NF-κB target binding sequence has been introduced into cells, usually with a liposome-facilitated internalization, to act as a decoy for active NF-κB moieties. This "decoy" approach could be rationally expected to provide as much specificity as is inherent in the transcription factor itself. In this regard, it should be noted that several native κB elements have been demonstrated to bind additional proteins, either through crossover of sequence affinities or as tandem sites that contain dual targets. In the latter, NF-κB sometimes appears to engage in tripartite binding to DNA the other protein, which is simultaneously bound to adjacent elements of the DNA [61-64]. One of the most widely used sequences for early "decoy" experiments is that of the HIV and immunoglobulin κ light chain promoters: GGGGACTTTCCCAGGC. In addition to NF-κB, this sequence readily binds RBP-Jκ and several members of the Sp1 family [65].

Genetic ablation ("knockout") of NF-κB subunits has interesting outcomes that may be revealing about their individual roles in various processes. RelA knockout mice die at embryonic day 14 or 15 due to hepatotoxicity triggered by TNFα [66]. By combining RelA knockout with knockout of TNFα (or TNF receptor 1), mice can be successfully obtained that are developmentally normal but have deficits in immune/inflammatory responses. By contrast, genetic manipulation to remove p50 by ablation of its p105 precursor protein has no dramatic effects on development. Perhaps this results from compensation by p52, but this seems unlikely given the obviously different roles of canonical versus noncanonical NF-κB in development [67]. It is possible that p50-containing moieties (such as the canonical NF-κB's RelA/p50 heterodimer) are less important for normal development and function of the CNS than are RelA homodimers and other forms of NF-κB. At any rate, the dramatic difference in outcomes after knockout of RelA versus p105/p50 hints at a rather limited utility of the latter. As suggested above, the genetic ablation of p105 would be expected to have complicated effects; p105 is itself acts as an IκB, and the removal of p50 would also relieve the system of inhibitory p50 homodimers. More sophisticated genetic models, such as cell-type restriction of expression [68], have begun to provide more rigorous data. Studies have also been performed in genetically altered models that exert conditional expression, particularly of a transdominant-repressor [69, 70].mutant of IκBα [69, 70]. However, even this approach has its caveats of specificity. IκBα inhibits not only NF-κB but also Sp1-related factors, for instance [71].

Pharmacological inhibition of of NF-κB has made tremendous advances since the isolation of IKKs and the development of inhibitors for these key components of NF-κB activation. Several compounds have been identified that provide relatively specific and potent inhibition of either multiple IKKs or IKK-2 (Table **1**). The key role of the latter in canonical NF-κB activation pathways makes its inhibitors particularly useful for blocking induction by proinflammatory stimuli or for discriminating between canonical and noncanonical pathways. To be sure, all of these later-generation inhibitors are preferable to the antioxidants and other first-generation "NF-κB inhibitors."

In addition to the empirical power of NF-κB removal or inhibition, insights may be gained through strategies that involve overexpression of NF-κB subunits. This approach is not foolproof; expression of NF-κB in an ectopic location does not reveal much about the normal role of the transcription factor. And because NF-κB activity is regulated by stoichiometric protein-protein interactions, overexpression of any of its subunits can result in aberrant activities due to disrupting the balance between transactivators and inhibitors. Thus, overexpression of transcriptional inducers (RelA, RelB, and c-Rel) is most effective as a means to achieve constitutive activation, independent of activation mechanisms that rely upon post-translational modification.

NF-κB IN NEURONS: EVIDENCE AND ERRORS

Several prominent reports have resulted in a conventional understanding that NF-κB is both 1) constitutively active in CNS neurons, and 2) inducible by excitatory neurotransmitters in CNS neurons [72-79]. All of these studies were conducted with brain cell cultures that varied in their proportions of neurons and glia. This is obviously a caveat when assay techniques are applied that utilize cell homogenates and lysates (e.g., EMSA, western blot, biochemical reporter gene assays), as glial proteins could easily contribute to the endpoint data. This is particularly important with an analyte such as NF-κB, which can undergo 100-fold changes in activity level under experimental conditions; as little as 5% glia could thereby contribute a five-fold change in endpoint readout. These scenarios have been further complicated by the failure to recognize important differences between post-mitotic neurons and cell lines used to model them. Together, these issues have led to an unfortunate degree of misunderstanding and misleading information.

Table 1.   IKK Inhibitors

4-amino-[2,3'-bithiophene]-5-carboxamide (SC-514)	IKK-2 (IC50 ~3-12 μM) IKK-1 (IC50 >200 μM)
2-Amino-6-(2-(cyclopropylmethoxy)-6-hydroxyphenyl)-4--(4-piperidinyl)-3-pyridinecarbonitrile (ACHP)	IKK-2 (IC50 =8.5 nM) IKK-1 (IC50 =250 nM)
4-(2′-Aminoethyl)amino-1,8-dimethylimidazo[1,2-a] quinoxaline (BMS-345541)	IKK-2 (IC50 ~300 nM) IKK-1 (IC50 ~4 μM)
N-(6-Chloro-9H-β-carbolin-8-yl)nicotinamide	IKK-1, -2 (IC50 =88 nM)
N-(3,5-Bis-trifluoromethylphenyl)-5-chloro-2-hydroxybenzamide (IMD-0354)	IKK-2 (IC50 ~250 nM)
5-(5,6-Dimethoxybenzimidazol-1-yl)-3-(2-methanesulfonyl-benzyloxy)-thiophene-2-carbonitrile	IKK-3 (IC50 =40 nM)
[5-(p-Fluorophenyl)-2-ureido]thiophene-3-carboxamide (TPCA-1)	IKK-2 (IC50 =18 nM)
7-Methoxy-5,11,12-trihydroxy-coumestan (Wedelolactone)	IKK-1, -2 (IC50 <10 μM)
6-oxo-3-(2-[4-(N-pyridin-2ylsulfamoyl)phenyl]hydrazono) cyclohexa-1,4-dienecarboxylic acid (sulfasalazine)	IKK-1, -2 (IC50 ~0.2 mM)
(5-Phenyl-2-ureido)thiophene-3-carboxamide	IKK-2 (IC50 =13 nM)
(5-Phenyl-3-ureido)thiophene-2-carboxamide	IKK-2 (IC50 =25 nM) IKK-1 (IC50 =1.0)

Because of an interest in both excitatory amino acid neurotransmission and neuroinflammation, the authors' laboratory made extensive attempts to repeat the reports of coupling between glutamate receptors and NF-κB. Following upon the precedent set by others, some experiments were performed with immunocytochemical detection of nuclear translocation in cultures of cerebral neurons. As shown in Fig. (**2A**), glutamate evoked a clear increase in the nuclear localization of RelA under these assay conditions, consistent with prior reports.

Early in this line of investigation we recognized that the contamination of brain cell cultures by glia would confound data interpretation. In conventional cell culture techniques, based on the classic methods developed by Banker *et al.* [80], serum-containing medium is used to maintain CNS neurons from regions of the fetal rodent brain. Subsequent refinements included the limitation of glial numbers by inclusion of mitotic inhibitors such as cytosine arabinoside (Ara-C) [81] or the substitution of serum by defined, quantified supplemental agents [82, 83]. Nevertheless, we found it was necessary to combine both serum-free medium and Ara-C treatment to restrict glial contamination below 2% of cells [84]. Such cultures become useful for studies of glutamate-receptor responses at 8-10 days *in vitro*, by which time glutamate receptor expression has matured [85].

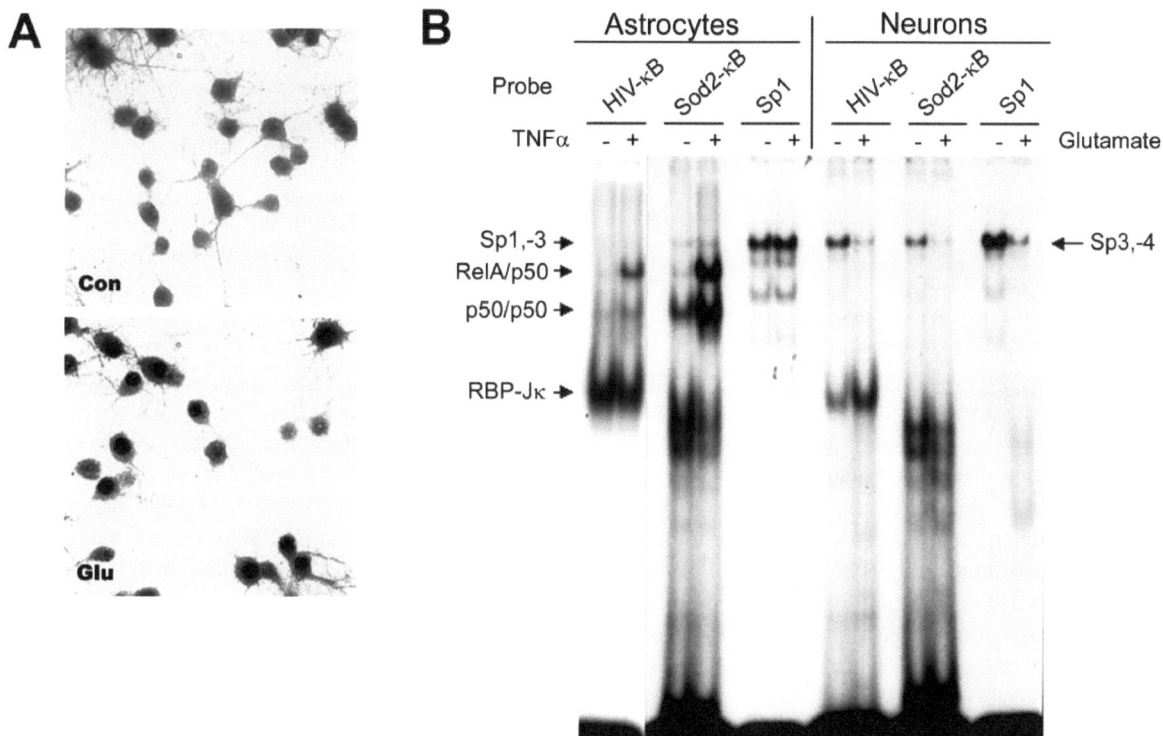

Fig. (2). Exploration of NF-κB activation in CNS neurons. A. Hippocampal neurons from rat were treated for 30 min with 50 μM glutamate, then fixed and subjected to immunocytochemistry for active RelA. Note the prominent staining of cell nuclei after treatment. **B.** DNA binding proteins were surveyed in nuclear extracts from nearly pure cultures of astrocytes and cortical neurons by EMSA. The astrocytes were treated with TNFα (100 ng/ml, 60 min) and neurons with glutamate (50 μM, 60 min). Three different DNA sequences were utilized as probes: the κB element from the HIV promoter, the κB element from the *Sod2* gene, and a Sp1 probe. Supershift assays (not shown) identified the lowest-mobility complex as comprising Sp1 and Sp3 in astrocytes and Sp3 and Sp4 in neurons [87], and these bound to all three probes. Supershifts also indicated that κB probes were bound by RelA/p50 and p50/p50 in astrocyte extracts only, as well as RBP-Jκ in both cell types. (The high-mobility complexes binding the Sod2-κB probe have not been identified.) Glutamate treatment of neurons did not induce NF-κB activity, though it substantially diminished the activity of Sp3 and -4.

After development of methods for obtaining nearly pure cultures of telencephalic neurons, nuclear translocation of NF-κB by glutamate or other glutamate receptor agonists was confirmed by western blot analysis after cell fractionation of both neocortical and hippocampal neurons. Nevertheless, glutamate was never observed to activate NF-κB as assayed by DNA binding or transcriptional activation [65, 84, 86-89]. Positive controls for reliable assay parameters included the assay of astrocytes and other cell types; e.g., activated with tumor necrosis factor (TNF)-α or lipopolysaccharide. Notably, neurons do contain nuclear proteins capable of binding certain κB enhancer sequences. These include Sp3 and Sp4 (members of the specific protein-1 family of transcription factors) and RBP-Jκ [84, 88-90]. Misidentification of these proteins may have contributed to the initial claims that "NF-κB" is constitutively active in CNS neurons; identification in early studies relied heavily on the mistaken assumption that κB sequences are exclusively bound by NF-κB. Fig. (**2B**) demonstrates a comparison of the κB binding factors detected by EMSA in relatively pure cultures of astrocytes and neurons.

Though initial attempts to activate neuronal NF-κB in this laboratory were focused on excitatory amino acid neurotransmitters, we have also tested other agonists. To date, NF-κB activity assays have failed in pure neuronal cultures stimulated with TNFα, nerve growth factor, brain-derived neurotrophic factor, cholinergic or adrenergic agonists, cytoskeletal drugs, or oxidants such as hydrogen peroxide (data not shown). Subsequently, others have confirmed a similar degree of recalcitrance of NF-κB in CNS neurons [91, 92].

In contrast to the nearly pure cultures of neurons, mixed neuron-glia cocultures have shown responses to many of the stimuli listed above. Even glutamate routinely stimulates such cocultures to show NF-κB DNA-binding activity by EMSA and reporter-gene assays [86]. Two approaches were applied to ascertain the cell type exhibiting NF-κB in such mixed cultures. First, cultures were established in which the glia were physically separated from the neurons in basket-type cell-culture inserts [87]. In this format, the cell types exchange soluble factors across a semipermeable membrane but can still be separated prior to lysis and assay. This paradigm demonstrated that glutamate activates NF-κB exclusively in glia, provided that the glia have been maintained for at least two days in the presence of neurons; naïve glia did not respond even when acutely placed in the presence of neurons.

The lack of NF-κB activation in naïve glia points out an important caveat: negative results in pure cultures of glia cannot be used to exclude their contributions to NF-κB in mixed cultures, as their responses are modified by prolonged exposure to neurons. A similar requirement for trans-cellular synergy might have explained the recalcitrance of NF-κB in neurons. Influences of glia on neurophysiology are many and include fundamental effects on neurotransmitter responses that are dependent upon direct contact between the two cell types [93]. To test whether contact with glia alters the responses of NF-κB in neurons, we utilized a mouse line that is transgenic for β-galactosidase driven by a promoter containing κB enhancer elements. This construct has been shown to report NF-κB activity through increases in β-galactosidase expression [94], though it is not clear how specifically the promoter excludes influences from other factors (below). We generated mixed-glial cultures (primarily astrocytes) from wild-type mice, then plated telencephalic neurons from the κB/βgal-transgenic mice onto these. The existing glia appeared to squelch the proliferation of transgenic glia from the dissociated E17 brain tissue, as β-gal staining was not apparent in cells staining with markers of astrocytes or microglia (not shown). Neurons, on the other hand, efficiently seeded onto the astrocyte lawn, similar to their application in established culture models [95]. In this paradigm, neurons are in direct contact with astrocytes, yet the readout of β-gal activity is almost exclusively from the neurons. For comparison, other cultures were prepared in which the neurons were plated alone and subjected to our serum-free/ara-C treatment to achieve nearly complete absence of glia. Each of these culture types was subjected to one of two glutamate treatments: 1) a transient exposure of 10 minutes, followed by a 12-hour chase, or 2) constant exposure to a lower concentration of glutamate for 12 hours. As shown in Table **2**, neither glutamate treatment elevated expression from the κB-driven promoter. In fact, glutamate diminished expression, consistent with its effect on Sp1-related transcription factors that also bind certain κB promoters [84, 86, 89]. Similar results were obtained at treatment and/or chase times of 6 hours and 24 hours. The presence of glia attenuated this effect of glutamate, probably by efficient glutamate transport and other trophic influences that astrocytes exert on neuronal health.

Table 2. Suppression of κB/βgal expression by glutamate

Treatment	β-gal activity (% control)	
	+ glia	- glia
untreated	100 ± 14.4	100 ± 3.10
transient Glu (50 μM)	42.6 ± 2.10*	4.1 ± 4.30**
continuous Glu (20 μM)	40.8 ± 1.30*	2.8 ± 0.70**

*p<0.02; **p<0.001

Questionable methodology continues to plague the field. Each generation of researcher seems destined to discover for themselves the difficulties in achieving highly enriched neuronal cultures. A 2009 article [96] reported NF-κB to be involved in induction of the NOX2 subunit of NADPH oxidase by interleukin-6 in "neuronal cultures"; but the prior papers cited for culture methods state that the cells were maintained in a mixture of fetal bovine serum and horse serum until "glia has reached confluency" [sic]. This paper also failed to measure transcription—or, for that matter, mRNA—of the gene supposedly activated by NF-κB, relying on immunocytochemistry of NOX2 protein as an endpoint. Finally, the report characterized its implied NF-κB activity as "noncanonical" without any demonstration of such. Misplaced confidence in cell culture purity is not limited to NF-κB neophytes either. Baltimore and colleagues were among the first to describe NF-κB and have done as much as any single laboratory to elucidate key components of its activation and function. However, this expertise did not render the laboratory immune to the artifacts presented by glial contamination of ostensibly neuronal cultures [79]. Even their use of isolated synaptosomes failed to recognize the well-established fact that such preparations contain significant contributions from astrocytic processes [97, 98]

Another approach prone to artifacts also tripped up Meffert *et al.*: overexpression of NF-κB subunits through transfection. It is quite clear that overexpression of RelA generates constitutively active homodimers [99]. It is also clear that this results from stoichometrically overwhelming the endogenous IκBs. Regardless of the care taken to titrate transfections to achieve low expression, adding something that is not normally present, expressed from a promoter that contains none of the normal autoregulatory capacity, is likely to generate artifact.

One of the biggest sources of confusion about the neuroscience of NF-κB is the utilization of neuroblastoma cells and other immortalized cell lines. PC12 cells are not neurons, yet prominent papers have swayed scientific opinion with the analysis of NF-κB regulation in these tumor cells and the claim that results from these cells tell us something about "neurons" [100]. Likewise, SH-SY5Y cells and related lines are highly proliferative, contain very unstable genomes, and do not exhibit excitotoxicity. NF-κB appears to be most active in actively mitotic cells, including neuroprogenitor cells [101]. Indeed, some tumor suppressor genes such as CYLD appear to work, at least in part, by inhibiting NF-κB and thereby blocking the transcription factor's anti-apoptotic actions [102]. This may have relevance to catastrophic mitosis, a mechanism that suppresses neoplasia by triggering apoptosis in cells that have initiated inappropriate mitotic overtures (e.g., mutagenic activation of a single oncogene). The fact that c-Rel, Bcl-3, and other genes associated with NF-κB activation were initially discovered as oncogenes is suggestive of a role for this class of transcription factors in promoting cell cycle progression; IκBα genes are deleted in Hodgkin's lymphoma, and IKKs are constitutively activated in melanomas and breast cancer [103, 104]. In general, NF-κB exerts anti-apoptotic effects [105-107]. Thus, one of its important roles may be in permitting progression of the cell cycle past apoptotic checkpoints associated with catastrophic mitosis, similar to the role in hyperplasia proposed for Bcl-2. Such activity would be unexpected in post-mitotic neurons, where catastrophic mitosis has been invoked to explain neuronal death in a variety of neurodegenerative conditions [108]. Regardless of the relevance of this phenomenon to NF-κB activity in neurons, there are many reasons to expect that its regulation would be quite irregular in tumor cells, including immortalized "neuronal" cell lines.

In vivo studies of NF-κB in the CNS have relied on both functional outcomes and transgenic reporter-gene. Data from two of the reporter-gene models have been used to conclude that NF-κB is constitutively active in neurons [94, 109]. Bhakar *et al.* reported that constitutive NF-κB activity could be detected by the κB/βgal reporter transgene, but EMSAs failed to demonstrate DNA-binding activity for NF-κB, which raises the possibility that the basal activity of this κB/βgal reporter is driven by other κB-binding factor(s), such as Sp1-related factors. This would be consistent with the decrease in βgal expression caused by glutamate of these transgenic neurons (Table **2**). Fridmacher *et al.* [109] used a reporter gene driven by the p105 promoter. This promoter was selected because it is NF-κB-responsive, but it also responds to Sp1-related factors binding to κB and non-κB elements [110]. Indeed, a band consistent with Sp1-related factors was prominently detected by EMSA in the initial report of this p105-promoter/βgal line [111]. The authors applied to this system a transdominant IκBα mutant (above). This manipulation did, in fact, inhibit expression of the p105-promoter/βgal reporter; but IκBα also inhibits activity of Sp1-related factors [71]. Fridmacher *et al.* did not report NF-κB in neurons by EMSA supershift or other immunological methods. Given all these caveats, it is not clear that these reports strongly support the claim that NF-κB is constitutively active in neurons.

In vivo studies arguing for a functional effect of NF-κB activation or inhibition have generally suffered from the same difficulties posed by mixed-cell cultures. Namely, it is difficult to ensure that the experimental endpoint reflects activity in neurons versus glia or even vascular elements. Expression of a transrepressor mutant of IκBα in mice has been used to argue that NF-κB is required for normal levels of neuronal viability and spatial memory. The mice generated by Fridmacher *et al.* ([109]; above) were utilized in studies that included expression of a transrepressor IκBα with a system that relies on elements of the calmodulin kinase IIα gene promoter. This promoter construct has been characterized as one that effects expression in the forebrain [112]. Mice expressing the transrepressor IκBα in this system showed a complete loss of NF-κB DNA-binding activity in the hippocampus, and hippocampal slice cultures generated from these mice accumulated greater degrees of cell death after being challenged with either kainic acid or ferrous sulfate [109]. In addition to the caveats noted above for this model, it is unexpected that the transrepressor expression would inhibit all NF-κB activity if its expression were restricted to neurons; glia appear responsible for a substantial fraction of NF-κB activity in the forebrain [101]. Indeed, inhibition of NF-κB specifically in astrocytes through this transrepressor IκBα strategy (utilizing the promoter of glial fibrillary acidic protein) was found to be sufficient to interfere with long-term potentiation (LTP) and spatial memory [70]. Thus, glial expression might explain the results of a study [69] showing memory deficits in the mouse line developed by Fridmacher *et al.*

The above caveats notwithstanding, it is possible that some type of activation of NF-κB, or merely its upstream signaling pathway, participates in learning or memory. Freudenthal *et al.* [113] reported activation of NF-κB, as measured by immunological detection of the unmasked RelA NLS (above) in what were ostensibly neuronal layers. Relatedly, application of κB decoy DNA to mouse hippocampal slices was found to block long-term depression and reduce the amplitude of LTP [114]. These findings extend to mammals intriguing data generated in the crab *Chasmagnathus* [115, 116]. Initial studies in this model were correlative, showing an increase in NF-κB activity by EMSA after a learning paradigm. A more rigorous association was afforded by using sulfasalazine to inhibit IKKs. This manipulation interfered with context-signal memory in the crabs. Of course, the allure of hypotheses that incorporate a transcription factor in mechanisms of learning and memory rely on the imagination that it might convey a signal from the synapse to the nucleus, therein inducing genes that somehow consolidate the synaptic change. Forgoing the caveat that it would be difficult to target such a nuclear event to the appropriate individual synapse, there is an additional impediment to this idea: to date, no NF-κB target gene has been identified that provides an obvious rational link to synaptic efficacy.

The mechanisms that might be responsible for NF-κB proteins achieving only partial activation in neurons are currently under investigation. To date, differential phosphorylation or acetylation of RelA in neurons has been excluded to the authors' satisfaction. Specifically, mutants of RelA that lack acetylation sites or Serine-276 (a phosphorylation site required in some systems) functioned as well as the wild-type RelA after transfection into primary neurons (data not shown). In addition, phosphorylation generally impacts transactivation rather than DNA binding (above), yet it is DNA binding that constitutes the first step in NF-κB activation where neurons show deficiency. Methodological artifacts arising from the use of progesterone and AraC in the culture medium were ruled out, as well. CNS neurons maintained in progesterone-free medium showed no greater capacity for NF-κB activation, and AraC has been washed out after the eradication of glia [99]. The inhibition of NF-κB by p53 does not appear to contribute, as the p53 inhibitor pifithrin-α had no effect on NF-κB reporter-gene activity in cultured neurons. There do not appear to be passive deficiencies in components of the NF-κB activation pathway, as their overexpression is ineffective (Fig. **3**); similar findings were obtained for the transcription coordinator p300. It is possible to observe robust DNA binding and reporter-gene induction in CNS neurons transfected with RelA (Fig. **3**), suggesting that if there is active inhibition of NF-κB activity, it can be overcome stoichiometrically. Such balance can also be restored by cotransfection of IκBα along with the RelA (not shown). However, western blot analysis indicates that expression levels of the major known IκBs are not inordinately high in CNS neurons. If anything, these cells trend toward a deficiency, as IκBα appears to be the only major IκB expressed at detectable levels in these cells (not shown).

For both the canonical and noncanonical heterodimers (RelA/p50 and RelB/p52), it is the "Rel" component exclusively that elicits transcriptional activation. Neither p50 nor p52 contains a transactivation domain; indeed, homodimers of such act as competitive inhibitors of NF-κB activity [9]. Similar to p52, p50 is generated through proteolytic processing of a precursor (p105), though this seems to happen constitutively for the pool that is destined for the canonical complex. Although p50 homodimers inhibit gene induction by transactivating dimers, Bcl-3 can act as an adapter protein, binding p50 homodimers and bridging them to the transcriptosome [10]. In addition, p50 has been found to heterodimerize with C/EBP proteins to augment transcription from a hybrid site that contains both a CCAAT element and a p50 "half site" [11]

The evidence that NF-κB activity is absent or, at least, drastically restricted in telencephalic cultures *in vitro* does not definitively rule out a role for this transcription factor in CNS neurons. However, there are rational arguments as to why conventional transcriptional activation by NF-κB might be repressed in these cells. First, as discussed by Massa *et al.* [65, 91], NF-κB often triggers expression of class I major histocompatibility complex proteins (MHC-

I). This is a key determinant for attack by cytotoxic T-cells during viral infection. Because of their finite numbers and critical contributions to life, CNS neurons would be a particularly maladaptive cell to target for removal. Thus, a suppression of NF-κB in neurons may be invoked to protect neurons from T-cell attack, a phenomenon that may contribute to the persistence of viral infections in the CNS. A second argument for the restriction of NF-κB in neurons relates to their postmitotic state. NF-κB is critically involved in cell cycle regulation, generally evoking the expression of promitotic genes [117-119]. Not only would expression of these cell-cycle genes be unnecessary in postmitotic neurons, there is evidence that such ectoptic expression of mitotic proteins triggers neuronal apoptosis [120].

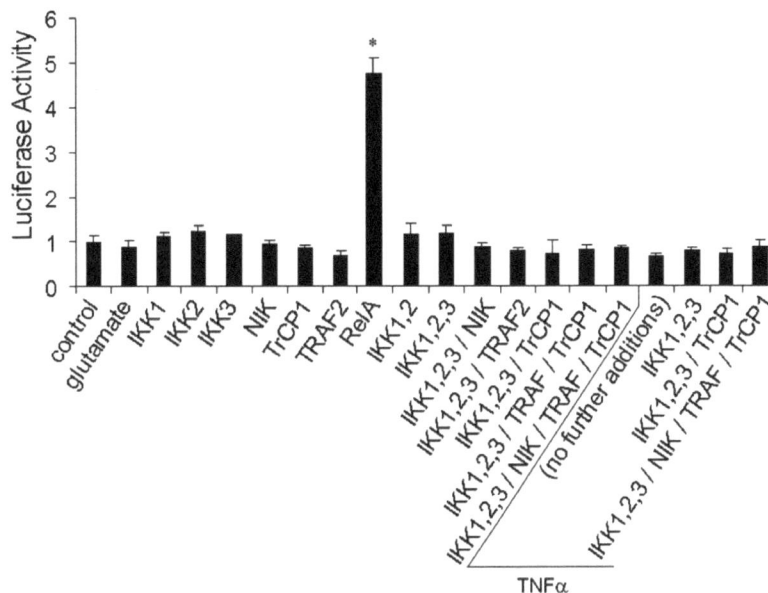

Fig. (3). Attempts to reconstitute NF-κB activation in CNS neurons. Nearly pure cultures of rat cortical neurons were transfected with a luciferase reporter gene construct. RelA, NIK, IKK1, IKK2, IKK3, β-TrCP1, or TRAF2 were expressed from vectors cotransfected either alone or in combinations as indicated. Some cultures were additionally treated with glutamate (20 μM) or rat TNFα (10 ng/ml). After 24 h, firefly luciferase activities were determined relative to Renilla luciferase reference. Values represent mean ± SEM of triplicate cultures. RelA expression differed significantly (p<0.02, ANOVA and post hoc test) from all other conditions, which did not differ from one another.

Regardless of how the final score is settled regarding NF-κB in brain neurons, there is considerable evidence mounting for a role for this transcription factor in peripheral neurons. Superior cervical ganglion neurons and nodose sensory neurons appear to undergo regulation of their neurite-outgrowth activity *via* NF-κB regulation [14, 34]. Though the evidence is compelling that NF-κB participates in these events, it is not clear that this occurs through transcriptional activation. No genes have been identified that respond to NF-κB and influence neurite extension in these systems. In fact, a viral protein has been shown to utilize NF-κB in a scheme to inactivate p53 *via* a mechanism that does not require NF-κB-dependent transcription [121]. There are intriguing interactions between NF-κB and other nuclear proteins, such as glucocorticoid receptors [52], suggesting that NF-κB may at times function as a transcriptional corepressor or costimulator independently of binding DNA itself. NF-κB family members also interact with components of the cytoskeleton [122, 123]. It is possible that protein-protein interactions between NF-κB components and cytoskeletal proteins explain their influence on neurite outgrowth. Finally, the signal transduction events that typically lead to NF-κB activation may produce misleading data by branching off into unconventional signaling events. For example, in response to DNA damage, IKK3 has been reported to bypass the IKK active complex and IκBs, instead coordinating a cellular response with the ataxia telangiectasia-mutated (ATM) kinase [124]. IKK1 can phosphorylate histone H3, which enhances transcription at some genes [125, 126]; it also phosphorylates RelA that is already bound to DNA-bound RelA, resulting in repression of transcription at genes otherwise induced by NF-κB [2]. IKK2 also phosphorylates and inactivates the transcription factor FOXO3a during tumorigenesis [127], and IKK1 participates in keratinocyte differentiation in an NF-κB-independent manner [128].

These novel roles for the IKKs suggest that experiments that utilize their natural agonists or their pharmacological inhibitors could generate results that might be incorrectly assumed to involve NF-κB.

NF-κB INHIBITION—PROSPECTS FOR THERAPY

NF-κB has been targeted for inhibition by some therapeutic strategies because of its importance for expression of cytokines and other factors related to inflammation. As mentioned above, the actions of some PPARγ agonists appear to involve NF-κB inhibition, and aspirin probably owes some of its effectiveness to this effect as well [129, 130]. It was therapeutic objectives that first inspired the development of specific IKK inhibitors; of course, this was also advantageous to research objectives. NF-κB has anti-apoptotic actions in most cell types, leading to the speculation that inhibition of its activity—perhaps mistakenly interpreted as constitutive—would be dangerous to CNS neurons [120]. However, the preponderance of data suggests that quelling inflammation by inhibition of NF-κB will not cause significant levels of neuronal cell death. Salicylates and IKK inhibitors promote positive outcomes in a variety of conditions that would otherwise generate neurotoxicity [131-135]. Indeed, genetic ablation of IKK1 or IKK2 protects against certain forms of neuropathology while creating no apparent deficiencies in the CNS [136, 137]. Thus, neurons probably do not depend upon tonic NF-κB activity, an interpretation also supported by knockout of various NF-κB subunits. Pharmacotherapeutic inhibition of NF-κB for neurological conditions should therefore be achievable without significant problems for CNS health and function.

ACKNOWLEDGEMENTS

This work was supported by funds from the NIH (R01NS046439 and P01AG12411). The κB-βgal transgenic mice were a gift from Philip A. Barker (McGill Univ., Montreal).

REFERENCES

[1]   Senftleben U, Cao Y, Xiao G, Greten FR, Krahn G, Bonizzi G, *et al.* Activation by IKKalpha of a second, evolutionary conserved, NF-kappa B signaling pathway. Science 2001; 293(5534): 1495-9.

[2]   Lawrence T, Bebien M, Liu GY, Nizet V, Karin M. IKKalpha limits macrophage NF-kappaB activation and contributes to the resolution of inflammation. Nature 2005; 434(7037): 1138-43.

[3]   Chen FE, Huang DB, Chen YQ, Ghosh G. Crystal structure of p50/p65 heterodimer of transcription factor NF-kappaB bound to DNA. Nature 1998; 391(6665): 410-3.

[4]   Bottero V, Rossi F, Samson M, Mari M, Hofman P, Peyron JF. Ikappa b-alpha, the NF-kappa B inhibitory subunit, interacts with ANT, the mitochondrial ATP/ADP translocator. J Biol Chem 2001; 276(24): 21317-24.

[5]   Sun SC, Ganchi PA, Ballard DW, Greene WC. NF-kB controls expression of inhibitor IkBa: evidence for an inducible autoregulatory pathway. Science 1993; 259: 1912-1915.

[6]   Beinke S, Ley SC. Functions of NF-kappaB1 and NF-kappaB2 in immune cell biology. Biochem J 2004; 382(Pt 2): 393-409.

[7]   Muller JR, Siebenlist U. Lymphotoxin beta receptor induces sequential activation of distinct NF-kappa B factors *via* separate signaling pathways. J Biol Chem 2003; 278(14): 12006-12.

[8]   Saccani S, Pantano S, Natoli G. Modulation of NF-kappaB activity by exchange of dimers. Mol Cell 2003; 11(6): 1563-74.

[9]   Franzoso G, Bours V, Park S, Tomita-Yamaguchi M, Kelly K, Siebenlist U. The candidate oncoprotein Bcl-3 is an antagonist of p50/NF-kappa B-mediated inhibition. Nature 1992; 359(6393): 339-42.

[10]   Dechend R, Hirano F, Lehmann K, *et al.* The Bcl-3 oncoprotein acts as a bridging factor between NF-kappaB/Rel and nuclear co-regulators. Oncogene 1999; 18(22): 3316-23.

[11]   Agrawal A, Cha-Molstad H, Samols D, Kushner I. Transactivation of C-reactive protein by IL-6 requires synergistic interaction of CCAAT/enhancer binding protein beta (C/EBP beta) and Rel p50. J Immunol 2001; 166(4): 2378-84.

[12]   Imbert V, Rupec RA, Livolsi A, Pahl HL, Traenckner EB, Mueller-Dieckmann C, *et al.* Tyrosine phosphorylation of I kappa B-alpha activates NF-kappa B without proteolytic degradation of I kappa B-alpha. Cell 1996; 86(5): 787-98.

[13]   Takada Y, Mukhopadhyay A, Kundu GC, Mahabeleshwar GH, Singh S, Aggarwal BB. Hydrogen peroxide activates NF-kappa B through tyrosine phosphorylation of I kappa B alpha and serine phosphorylation of p65: evidence for the involvement of I kappa B alpha kinase and Syk protein-tyrosine kinase. J Biol Chem 2003; 278(26): 24233-41.

[14]   Gallagher D, Gutierrez H, Gavalda N, O'Keeffe G, Hay R, Davies AM. Nuclear factor-kappaB activation *via* tyrosine phosphorylation of inhibitor kappaB-alpha is crucial for ciliary neurotrophic factor-promoted neurite growth from developing neurons. J Neurosci 2007; 27(36): 9664-9.

[15]   Bui NT, Livolsi A, Peyron JF, Prehn JH. Activation of nuclear factor kappaB and Bcl-x survival gene expression by nerve growth factor requires tyrosine phosphorylation of IkappaBalpha. J Cell Biol 2001; 152(4): 753-64.

[16]   Qin Z, Wang Y, Chasea TN. A caspase-3-like protease is involved in NF-kappaB activation induced by stimulation of N-methyl-D-aspartate receptors in rat striatum. Brain Res Mol Brain Res 2000; 80(2): 111-22.

[17]   Waterfield MR, Zhang M, Norman LP, Sun SC. NF-kappaB1/p105 regulates lipopolysaccharide-stimulated MAP kinase signaling by governing the stability and function of the Tpl2 kinase. Mol Cell 2003; 11(3): 685-94.

[18]   Inoue J, Kerr LD, Kakizuka A, Verma IM. I kappa B gamma, a 70 kd protein identical to the C-terminal half of p110 NF-kappa B: a new member of the I kappa B family. Cell 1992; 68(6): 1109-20.

[19]   He B, Weber GF. Phosphorylation of NF-kappaB proteins by cyclic GMP-dependent kinase. A noncanonical pathway to NF-kappaB activation. Eur J Biochem 2003; 270(10): 2174-85.

[20]   Zhong H, SuYang H, Erdjument-Bromage H, Tempst P, Ghosh S. The transcriptional activity of NF-kappaB is regulated by the IkappaB- associated PKAc subunit through a cyclic AMP-independent mechanism. Cell 1997; 89(3): 413-24.

[21]   Zhong H, Voll RE, Ghosh S. Phosphorylation of NF-kappa B p65 by PKA stimulates transcriptional activity by promoting a novel bivalent interaction with the coactivator CBP/p300. Mol Cell 1998; 1(5): 661-71.

[22]   Anrather J, Csizmadia V, Soares MP, Winkler H. Regulation of NF-kappaB RelA phosphorylation and transcriptional activity by p21(ras) and protein kinase Czeta in primary endothelial cells. J Biol Chem 1999; 274(19): 13594-603.

[23]   Martin AG, Fresno M. Tumor necrosis factor-alpha activation of NF-kappa B requires the phosphorylation of Ser-471 in the transactivation domain of c-Rel. J Biol Chem 2000; 275(32): 24383-91.

[24]   Martin AG, San-Antonio B, Fresno M. Regulation of nuclear factor kappa B transactivation. Implication of phosphatidylinositol 3-kinase and protein kinase C zeta in c-Rel activation by tumor necrosis factor alpha. J Biol Chem 2001; 276(19): 15840-9.

[25]   Bird TA, Schooley K, Dower SK, Hagen H, Virca GD. Activation of nuclear transcription factor NF-kappaB by interleukin-1 is accompanied by casein kinase II-mediated phosphorylation of the p65 subunit. J Biol Chem 1997; 272(51): 32606-12.

[26]   Wang D, Westerheide SD, Hanson JL, Baldwin AS, Jr. Tumor necrosis factor alpha-induced phosphorylation of RelA/p65 on Ser529 is controlled by casein kinase II. J Biol Chem 2000; 275(42): 32592-7.

[27]   Wang D, Baldwin AS, Jr. Activation of nuclear factor-kappaB-dependent transcription by tumor necrosis factor-alpha is mediated through phosphorylation of RelA/p65 on serine 529. J Biol Chem 1998; 273(45): 29411-6.

[28]   Mercurio F, Zhu H, Murray BW, S *et al.* IKK-1 and IKK-2: cytokine-activated IkappaB kinases essential for NF-kappaB activation. Science 1997; 278(5339): 860-6.

[29]   Sakurai H, Chiba H, Miyoshi H, Sugita T, Toriumi W. IkappaB kinases phosphorylate NF-kappaB p65 subunit on serine 536 in the transactivation domain. J Biol Chem 1999; 274(43): 30353-6.

[30]   Sizemore N, Lerner N, Dombrowski N, Sakurai H, Stark GR. Distinct roles of the Ikappa B kinase alpha and beta subunits in liberating nuclear factor kappa B (NF-kappa B) from Ikappa B and in phosphorylating the p65 subunit of NF-kappa B. J Biol Chem 2002; 277(6): 3863-9.

[31]   Yang F, Tang E, Guan K, Wang CY. IKK beta plays an essential role in the phosphorylation of RelA/p65 on serine 536 induced by lipopolysaccharide. J Immunol 2003; 170(11): 5630-5.

[32]   Jang MK, Goo YH, Sohn YC, *et al.* Ca^{2+}/calmodulin-dependent protein kinase IV stimulates nuclear factor-kappa B transactivation *via* phosphorylation of the p65 subunit. J Biol Chem 2001; 276(23): 20005-10.

[33]   Schwabe RF, Brenner DA. Role of glycogen synthase kinase-3 in TNF-alpha-induced NF-kappaB activation and apoptosis in hepatocytes. Am J Physiol Gastrointest Liver Physiol 2002; 283(1): G204-11.

[34]   Gutierrez H, O'Keeffe GW, Gavalda N, Gallagher D, Davies AM. Nuclear factor kappa B signaling either stimulates or inhibits neurite growth depending on the phosphorylation status of p65/RelA. J Neurosci 2008; 28(33): 8246-56.

[35]   Sasaki CY, Barberi TJ, Ghosh P, Longo DL. Phosphorylation of RelA/p65 on serine 536 defines an IkappaBalpha-independent NF-kappaB pathway. J Biol Chem 2005; 280(41): 34538-47.

[36]   Chen L, Fischle W, Verdin E, Greene WC. Duration of nuclear NF-kappaB action regulated by reversible acetylation. Science 2001; 293(5535): 1653-7.

[37]   Chen LF, Mu Y, Greene WC. Acetylation of RelA at discrete sites regulates distinct nuclear functions of NF-kappaB. Embo J 2002; 21(23): 6539-48.

[38]   Chen LF, Williams SA, Mu Y, *et al.* NF-kappaB RelA phosphorylation regulates RelA acetylation. Mol Cell Biol 2005; 25(18): 7966-75.

[39]   Kiernan R, Bres V, Ng RW, *et al.* Post-activation turn-off of NF-kappa B-dependent transcription is regulated by acetylation of p65. J Biol Chem 2003; 278(4): 2758-66.

[40]   Yeung F, Hoberg JE, Ramsey CS, *et al.* Modulation of NF-kappaB-dependent transcription and cell survival by the SIRT1 deacetylase. Embo J 2004; 23(12): 2369-80.

[41]   Buerki C, Rothgiesser KM, Valovka T, *et al.* Functional relevance of novel p300-mediated lysine 314 and 315 acetylation of RelA/p65. Nucleic Acids Res 2008; 36(5): 1665-80.

[42]   Kaltschmidt C, Kaltschmidt B, Henkel T, *et al.* Selective recognition of the activated form of transcription factor NF-kappa B by a monoclonal antibody. Biol Chem Hoppe Seyler 1995; 376(1): 9-16.

[43]   Mukaida N, Morita M, Ishikawa Y, Rice N, Okamoto S, Kasahara T. Novel mechanism of glucocorticoid-mediated gene repression. Nuclear factor-kB is target for glucocorticoid-mediated interleukin 8 gene repression. J. Biol. Chem. 1994; 269: 13289-13295.

[44]   Brostjan C, Anrather J, Csizmadia V, *et al.* Glucocorticoid-mediated repression of NFkappaB activity in endothelial cells does not involve induction of IkappaBalpha synthesis. J Biol Chem 1996; 271(32): 19612-6.

[45]   Ray KP, Farrow S, Daly M, Talabot F, Searle N. Induction of the E-selectin promoter by interleukin 1 and tumour necrosis factor alpha, and inhibition by glucocorticoids. Biochem J 1997; 328 (Pt 2): 707-15.

[46]   Harant H, Wolff B, Lindley IJ. 1Alpha,25-dihydroxyvitamin D3 decreases DNA binding of nuclear factor-kappaB in human fibroblasts. FEBS Lett 1998; 436(3): 329-34.

[47]   Liden J, Rafter I, Truss M, Gustafsson JA, Okret S. Glucocorticoid effects on NF-kappaB binding in the transcription of the ICAM-1 gene. Biochem Biophys Res Commun 2000; 273(3): 1008-14.

[48]   True AL, Rahman A, Malik AB. Activation of NF-kappaB induced by H(2)O(2) and TNF-alpha and its effects on ICAM-1 expression in endothelial cells. Am J Physiol Lung Cell Mol Physiol 2000; 279(2): L302-11.

[49]   Leitges M, Sanz L, Martin P, *et al.* Targeted disruption of the zetaPKC gene results in the impairment of the NF-kappaB pathway. Mol Cell 2001; 8(4): 771-80.

[50]   Din FV, Stark LA, Dunlop MG. Aspirin-induced nuclear translocation of NFkappaB and apoptosis in colorectal cancer is independent of p53 status and DNA mismatch repair proficiency. Br J Cancer 2005; 92(6): 1137-43.

[51]   Liu GH, Qu J, Shen X. NF-kappaB/p65 antagonizes Nrf2-ARE pathway by depriving CBP from Nrf2 and facilitating recruitment of HDAC3 to MafK. Biochim Biophys Acta 2008; 1783(5): 713-27.

[52]　Ray A, Prefontaine KE. Physical association and functional antagonism between the p65 subunit of transcription factor NF-kB and the glucocorticoid receptor. Proc Natl Acad Sci USA 1994; 91: 752-756.

[53]　Kopp E, Ghosh S. Inhibition of NF-kB by sodium salicylate and aspirin. Science 1994; 265: 956-959.

[54]　Straus DS, Pascual G, Li M, *et al.* 15-deoxy-delta 12,14-prostaglandin J2 inhibits multiple steps in the NF-kappa B signaling pathway. Proc Natl Acad Sci U S A 2000; 97(9): 4844-9.

[55]　Castrillo A, Diaz-Guerra MJ, Hortelano S, Martin-Sanz P, Bosca L. Inhibition of IkappaB kinase and IkappaB phosphorylation by 15-deoxy-Delta(12,14)-prostaglandin J(2) in activated murine macrophages. Mol Cell Biol 2000; 20(5): 1692-8.

[56]　Rossi A, Kapahi P, Natoli G, *et al.* Anti-inflammatory cyclopentenone prostaglandins are direct inhibitors of IkappaB kinase. Nature 2000; 403(6765): 103-8.

[57]　Pereira MP, Hurtado O, Cardenas A, *et al.* Rosiglitazone and 15-deoxy-Delta12,14-prostaglandin J2 cause potent neuroprotection after experimental stroke through noncompletely overlapping mechanisms. J Cereb Blood Flow Metab 2006; 26(2): 218-29.

[58]　Hayakawa M, Miyashita H, Sakamoto I, *et al.* Evidence that reactive oxygen species do not mediate NF-kB activation. EMBO J. 2003; 22: 3356-3366.

[59]　Pikarsky E, Melamed E, Rosenthal J, Uzzan A, Michowiz SD. The neurotoxin MPTP does not affect striatal superoxide dismutase activity in mice. Neurosci Lett 1987; 82(3): 327-31.

[60]　Xu Y, Fang F, St Clair DK, Sompol P, Josson S, St Clair WH. SN52, a novel nuclear factor-kappaB inhibitor, blocks nuclear import of RelB: p52 dimer and sensitizes prostate cancer cells to ionizing radiation. Mol Cancer Ther 2008; 7(8): 2367-76.

[61]　Perkins ND, Agranoff AB, Pascal E, Nabel GJ. An interaction between the DNA-binding domains of RelA(p65) and Sp1 mediates human immunodeficiency virus gene activation. Mol Cell Biol 1994; 14(10): 6570-83.

[62]　Cakouros D, Cockerill PN, Bert AG, Mital R, Roberts DC, Shannon MF. A NF-kappa B/Sp1 region is essential for chromatin remodeling and correct transcription of a human granulocyte-macrophage colony-stimulating factor transgene. J Immunol 2001; 167(1): 302-10.

[63]　Stein B, Baldwin AS, Jr. Distinct mechanisms for regulation of the interleukin-8 gene involve synergism and cooperativity between C/EBP and NF-kappa B. Mol Cell Biol 1993; 13(11): 7191-8.

[64]　Marampon F, Casimiro MC, Fu M, *et al.* Nerve Growth factor regulation of cyclin D1 in PC12 cells through a p21RAS extracellular signal-regulated kinase pathway requires cooperative interactions between Sp1 and nuclear factor-kappaB. Mol Biol Cell 2008; 19(6): 2566-78.

[65]　Massa PE, Aleyasin H, Park DS, Mao X, Barger SW. NFkB in neurons? The uncertainty principle in neurobiology. J. Neurochem. 2006; 97(3): 607-618.

[66]　Doi TS, Marino MW, Takahashi T, *et al.* Absence of tumor necrosis factor rescues RelA-deficient mice from embryonic lethality. Proc Natl Acad Sci U S A 1999; 96(6): 2994-9.

[67]　Pasparakis M, Luedde T, Schmidt-Supprian M. Dissection of the NF-kappaB signalling cascade in transgenic and knockout mice. Cell Death Differ 2006; 13(5): 861-72.

[68]　Inta I, Paxian S, Maegele I, *et al.* Bim and Noxa are candidates to mediate the deleterious effect of the NF-kappa B subunit RelA in cerebral ischemia. J Neurosci 2006; 26(50): 12896-903.

[69]　Kaltschmidt B, Ndiaye D, Korte M, *et al.* NF-kappaB regulates spatial memory formation and synaptic plasticity through protein kinase A/CREB signaling. Mol Cell Biol 2006; 26(8): 2936-46.

[70]　Bracchi-Ricard V, Brambilla R, Levenson J, *et al.* Astroglial nuclear factor-kappaB regulates learning and memory and synaptic plasticity in female mice. J Neurochem 2008; 104(3): 611-23.

[71]　Heckman CA, Mehew JW, Boxer LM. NF-kappaB activates Bcl-2 expression in t(14; 18) lymphoma cells. Oncogene 2002; 21(24): 3898-908.

[72]　Kaltschmidt C, Kaltschmidt B, Baeuerle PA. Brain synapses contain inducible forms of the transcription factor NF-kB. Mech Dev 1993; 43: 135-147.

[73]　Kaltschmidt C, Kaltschmidt B, Neumann H, Wekerle H, Baeuerle PA. Constitutive NF-kB activity in neurons. Mol Cell Biol 1994; 14: 3981-3992.

[74]　Kaltschmidt C, Kaltschmidt B, Baeuerle PA. Stimulation of ionotropic glutamate receptors activates transcription factor NF-kB in primary neurons. Proc Natl Acad Sci USA 1995; 92: 9618-9622.

[75]　Guerrini L, Blasi F, Denis-Donini S. Synaptic activation of NF-kB by glutamate in cerebellar granule neurons in vitro. Proc Natl Acad Sci USA 1995; 92: 9077-9081.

[76]　Grilli M, Goffi F, Memo M, Spano P. Interleukin-1b and glutamate activate the NF-kB/Rel binding site from the regulatory region of the amyloid precursor protein gene in primary neuronal cultures. J Biol Chem 1996; 271: 15002-15007.

[77]　Grilli M, Pizzi M, Memo M, Spano P. Neuroprotection by aspirin and sodium salicylate through blockade of NF-kB activation. Science 1996; 274: 1383-1385.

[78]　Guerrini L, Molteni A, Wirth T, Kistler B, Blasi F. Glutamate-dependent activation of NF-kB during mouse cerebellum development. J. Neurosci 1997; 17: 6057-6063.

[79]　Meffert MK, Chang JM, Wiltgen BJ, Fanselow MS, Baltimore D. NF-kappa B functions in synaptic signaling and behavior. Nat Neurosci 2003; 6(10): 1072-8.

[80]　Banker GA, Cowan WM. Rat hippocampal neurons in dispersed cell culture. Brain Res 1977; 126(3): 397-42.

[81]　Oorschot DE, Jones DG. Non-neuronal cell proliferation in tissue culture: implications for axonal regeneration in the central nervous system. Brain Res 1986; 368(1): 49-61.

[82]　Bottenstein JE, Skaper SD, Varon SS, Sato GH. Selective survival of neurons from chick embryo sensory ganglionic dissociates utilizing serum-free supplemented medium. Exp Cell Res 1980; 125(1): 183-90.

[83]　Brewer GJ, Torricelli JR, Evege EK, Price PJ. Optimized survival of hippocampal neurons in B27-supplemented Neurobasal, a new serum-free medium combination. J Neurosci Res 1993; 35: 567-576.

[84]　Mao X, Moerman AM, Barger SW. Neuronal kB-binding factors consist of Sp1-related proteins: Functional implications for autoregulation of NR1 expression. J Biol Chem 2002; 277(47): 44911-44919.

[85]　Peterson C, Neal JH, Cotman CW. Development of N-methyl-D-aspartate excitotoxicity in cultured hippocampal neurons. Brain Res Dev Brain Res 1989; 48(2): 187-95.

[86]　Mao X, Moerman AM, Lucas MM, Barger SW. Inhibition of the activity of a neuronal kB-binding factor (NKBF) by glutamate. J Neurochem 1999; 73: 1851-1858.

[87]　Moerman AM, Mao X, Lucas MM, Barger SW. Characterization of a neuronal kB-binding factor distinct from NF-kB. Mol. Brain Res. 1999; 67: 303-315.

[88]   Mao X, Moerman-Herzog AM, Wang W, Barger SW. Differential transcriptional control of the superoxide dismutase-2 kappaB element in neurons and astrocytes. J Biol Chem 2006; 281(47): 35863-72.

[89]   Mao X, Yang SH, Simpkins JW, Barger SW. Glutamate receptor activation evokes calpain-mediated degradation of Sp3 and Sp4, the prominent Sp-family transcription factors in neurons. J Neurochem 2007; 100(5): 1300-14.

[90]   Barger SW, Moerman AM, Mao X. Molecular mechanisms of cytokine-induced neuroprotection: NFkappaB and neuroplasticity. Curr Pharm Des 2005; 11(8): 985-98.

[91]   Jarosinski KW, Whitney LW, Massa PT. Specific deficiency in nuclear factor-kB activation in neurons of the central nervous system. Lab Invest 2001; 81: 1275-1288.

[92]   Saha RN, Pahan K. Greater availability of NF-kB p65: p50 in glia than neurons: implications for neurodegenerative disorders. J Neurochem 2005; 94: S127.

[93]   Nagler K, Mauch DH, Pfrieger FW. Glia-derived signals induce synapse formation in neurones of the rat central nervous system. J Physiol 2001; 533(Pt 3): 665-79.

[94]   Bhakar AL, Tannis LL, Zeindler C, *et al.* Constitutive nuclear factor-kappa B activity is required for central neuron survival. J Neurosci 2002; 22(19): 8466-75.

[95]   Huettner JE, Baughman RW. Primary culture of identified neurons from the visual cortex of postnatal rats. J Neurosci 1986; 6(10): 3044-60.

[96]   Dugan LL, Ali SS, Shekhtman G, *et al.* IL-6 mediated degeneration of forebrain GABAergic interneurons and cognitive impairment in aged mice through activation of neuronal NADPH oxidase. PLoS One 2009; 4(5): e5518.

[97]   Henn FA, Anderson DJ, Rustad DG. Glial contamination of synaptosomal fractions. Brain Res 1976; 101(2): 341-4.

[98]   Chicurel ME, Terrian DM, Potter H. mRNA at the synapse: analysis of a synaptosomal preparation enriched in hippocampal dendritic spines. J Neurosci 1993; 13(9): 4054-63.

[99]   Mao XR, Moerman-Herzog AM, Chen Y, Barger SW. Unique aspects of transcriptional regulation in neurons--nuances in NFkappaB and Sp1-related factors. J Neuroinflammation 2009; 6: 16.

[100]   Sole C, Dolcet X, Segura MF, *et al.* The death receptor antagonist FAIM promotes neurite outgrowth by a mechanism that depends on ERK and NF-kapp B signaling. J Cell Biol 2004; 167(3): 479-92.

[101]   Denis-Donini S, Caprini A, Frassoni C, Grilli M. Members of the NF-kappaB family expressed in zones of active neurogenesis in the postnatal and adult mouse brain. Brain Res Dev Brain Res 2005; 154(1): 81-9.

[102]   Stegmeier F, Sowa ME, Nalepa G, Gygi SP, Harper JW, Elledge SJ. The tumor suppressor CYLD regulates entry into mitosis. Proc Natl Acad Sci USA 2007; 104(21): 8869-74.

[103]   Sun SC, Xiao G. Deregulation of NF-kappaB and its upstream kinases in cancer. Cancer Metastasis Rev 2003; 22(4): 405-22.

[104]   Li Q, Withoff S, Verma IM. Inflammation-associated cancer: NF-kappaB is the lynchpin. Trends Immunol 2005; 26(6): 318-25.

[105]   Van Antwerp DJ, Martin SJ, Kafri T, Green DR, Verma IM. Suppression of TNF-a-induced apoptosis by NF-kB. Science 1996; 274: 787-789.

[106]   Wang CY, Mayo MW, Baldwin ASJ. TNF- and cancer therapy-induced apoptosis: potentiation by inhibition of NF-kB. Science 1996; 274: 784-787.

[107]   Beg AA, Baltimore D. An essential role for NF-kB in preventing TNF-a-induced cell death. Science 1996; 274: 782-784.

[108]   Wang W, Bu B, Xie M, Zhang M, Yu Z, Tao D. Neural cell cycle dysregulation and central nervous system diseases. Prog Neurobiol 2009; 89(1): 1-17.

[109]   Fridmacher V, Kaltschmidt B, Goudeau B, *et al.* Forebrain-specific neuronal inhibition of nuclear factor-kappaB activity leads to loss of neuroprotection. J Neurosci 2003; 23(28): 9403-8.

[110]   Yurochko AD, Mayo MW, Poma EE, Baldwin AS, Jr., Huang ES. Induction of the transcription factor Sp1 during human cytomegalovirus infection mediates upregulation of the p65 and p105/p50 NF-kappaB promoters. J Virol 1997; 71(6): 4638-48.

[111]   Schmidt-Ullrich R, Memet S, Lilienbaum A, Feuillard J, Raphael M, Israel A. NF-kB activity in transgenic mice: developmental regulation and tissue specificity. Development 1996; 122: 2117-2128.

[112]   Mayford M, Bach ME, Huang YY, Wang L, Hawkins RD, Kandel ER. Control of memory formation through regulated expression of a CaMKII transgene. Science 1996; 274(5293): 1678-83.

[113]   Freudenthal R, Romano A, Routtenberg A. Transcription factor NF-kappaB activation after *in vivo* perforant path LTP in mouse hippocampus. Hippocampus 2004; 14(6): 677-83.

[114]   Albensi BC, Mattson MP. Evidence for the involvement of TNF and NF-kappaB in hippocampal synaptic plasticity. Synapse 2000; 35(2): 151-9.

[115]   Freudenthal R, Locatelli F, Hermitte G, *et al.* Kappa-B like DNA-binding activity is enhanced after spaced training that induces long-term memory in the crab Chasmagnathus. Neurosci Lett 1998; 242(3): 143-6.

[116]   Merlo E, Freudenthal R, Maldonado H, Romano A. Activation of the transcription factor NF-kappaB by retrieval is required for long-term memory reconsolidation. Learn Mem 2005; 12(1): 23-9.

[117]   Grumont RJ, Rourke IJ, O'Reilly LA, *et al.* B lymphocytes differentially use the Rel and nuclear factor kappaB1 (NF-kappaB1) transcription factors to regulate cell cycle progression and apoptosis in quiescent and mitogen-activated cells. J Exp Med 1998; 187(5): 663-74.

[118]   Tergaonkar V, Correa RG, Ikawa M, Verma IM. Distinct roles of IkappaB proteins in regulating constitutive NF-kappaB activity. Nat Cell Biol 2005; 7(9): 921-3.

[119]   Barre B, Perkins ND. A cell cycle regulatory network controlling NF-kappaB subunit activity and function. Embo J 2007; 26(23): 4841-55.

[120]   Yang Y, Herrup K. Cell division in the CNS: protective response or lethal event in post-mitotic neurons? Biochim Biophys Acta 2007; 1772(4): 457-66.

[121]   Jeong SJ, Pise-Masison CA, Radonovich MF, Park HU, Brady JN. A novel NF-kappaB pathway involving IKKbeta and p65/RelA Ser-536 phosphorylation results in p53 Inhibition in the absence of NF-kappaB transcriptional activity. J Biol Chem 2005; 280(11): 10326-32.

[122]   Are AF, Galkin VE, Pospelova TV, Pinaev GP. The p65/RelA subunit of NF-kappaB interacts with actin-containing structures. Exp Cell Res 2000; 256(2): 533-44.

[123]   Babakov VN, Petukhova OA, Turoverova LV, *et al.* RelA/NF-kappaB transcription factor associates with alpha-actinin-4. Exp Cell Res 2008; 314(5): 1030-8.

[124]   Huang TT, Wuerzberger-Davis SM, Wu ZH, Miyamoto S. Sequential modification of NEMO/IKKgamma by SUMO-1 and ubiquitin mediates NF-kappaB activation by genotoxic stress. Cell 2003; 115(5): 565-76.

[125]   Yamamoto Y, Verma UN, Prajapati S, Kwak YT, Gaynor RB. Histone H3 phosphorylation by IKK-alpha is critical for cytokine-induced gene expression. Nature 2003; 423(6940): 655-9.

[126]   Anest V, Hanson JL, Cogswell PC, Steinbrecher KA, Strahl BD, Baldwin AS. A nucleosomal function for IkappaB kinase-alpha in NF-kappaB-dependent gene expression. Nature 2003; 423(6940): 659-63.

[127]   Hu MC, Lee DF, Xia W, *et al.* IkappaB kinase promotes tumorigenesis through inhibition of forkhead FOXO3a. Cell 2004; 117(2): 225-37.

[128]   Hu Y, Baud V, Oga T, Kim KI, Yoshida K, Karin M. IKKalpha controls formation of the epidermis independently of NF-kappaB. Nature 2001; 410(6829): 710-4.

[129]   Bernardo A, Levi G, Minghetti L. Role of the peroxisome proliferator-activated receptor-gamma (PPAR-gamma) and its natural ligand 15-deoxy-Delta12, 14-prostaglandin J2 in the regulation of microglial functions. Eur J Neurosci 2000; 12(7): 2215-23.

[130]   Cernuda-Morollon E, Rodriguez-Pascual F, Klatt P, Lamas S, Perez-Sala D. PPAR agonists amplify iNOS expression while inhibiting NF-kappaB: implications for mesangial cell activation by cytokines. J Am Soc Nephrol 2002; 13(9): 2223-31.

[131]   Jiang H, Sha SH, Schacht J. NF-kappaB pathway protects cochlear hair cells from aminoglycoside-induced ototoxicity. J Neurosci Res 2005; 79(5): 644-51.

[132]   Zhao X, Zhang Y, Strong R, Grotta JC, Aronowski J. 15d-Prostaglandin J2 activates peroxisome proliferator-activated receptor-gamma, promotes expression of catalase, and reduces inflammation, behavioral dysfunction, and neuronal loss after intracerebral hemorrhage in rats. J Cereb Blood Flow Metab 2006; 26(6): 811-20.

[133]   Zheng L, Howell SJ, Hatala DA, Huang K, Kern TS. Salicylate-based anti-inflammatory drugs inhibit the early lesion of diabetic retinopathy. Diabetes 2007; 56(2): 337-45.

[134]   Kerr BJ, Girolami EI, Ghasemlou N, Jeong SY, David S. The protective effects of 15-deoxy-delta-(12,14)-prostaglandin J2 in spinal cord injury. Glia 2008; 56(4): 436-48.

[135]   Sarnico I, Boroni F, Benarese M, *et al.* Targeting IKK2 by pharmacological inhibitor AS602868 prevents excitotoxic injury to neurons and oligodendrocytes. J Neural Transm 2008; 115(5): 693-701.

[136]   Herrmann O, Baumann B, de Lorenzi R, *et al.* IKK mediates ischemia-induced neuronal death. Nat Med 2005; 11(12): 1322-9.

[137]   van Loo G, De Lorenzi R, Schmidt H, *et al.* Inhibition of transcription factor NF-kappaB in the central nervous system ameliorates autoimmune encephalomyelitis in mice. Nat Immunol 2006; 7(9): 954-61.

INDEX

www.ingramcontent.com/pod-product-compliance
Lightning Source LLC
Chambersburg PA
CBHW041715210326
41598CB00007B/656